Lecture Notes in Social Networks

More information about this series at http://www.springer.com/series/8768

Mohammad A. Tayebi • Uwe Glässer

Social Network Analysis in Predictive Policing

Concepts, Models and Methods

Mohammad A. Tayebi
Computing Science
Simon Fraser University
British Columbia, Canada

Uwe Glässer
Computing Science
Simon Fraser University
British Columbia, Canada

ISSN 2190-5428 ISSN 2190-5436 (electronic)
Lecture Notes in Social Networks
ISBN 978-3-319-82368-3 ISBN 978-3-319-41492-8 (eBook)
DOI 10.1007/978-3-319-41492-8

Printed on acid-free paper

This Springer imprint is published by Springer Nature
The registered company is Springer International Publishing AG Switzerland

Foreword

Policing resources across North America have become increasingly under pressure, and police governance authorities and governments are struggling to meet the increasing demands of both frontline policing and the complicated financial and social impacts of organized crime on society. Along with these pressures, the world of intelligence gathering has remained relatively stable and consistent in its use of human source information to inform law enforcement authorities on the location and proliferation of organized crime activities in our societies. The research demonstrated in this text shows an alternative evidence-based approach to the standard intelligence gathering process by enhancing law enforcement's preventative capacity in identifying organized crime groups that previously went undetected under standard police intelligence gathering techniques. The utilization of co-offending networks and geographical analysis provides an unbiased scientific methodology to the intelligence process that in addition to human source techniques increases the productivity and accountability of policing resources in the detection and strength of organized crime groups. Early identification and detection of these groups through predictive policing ensures that both law enforcement and communities can proactively engage and mobilize community efforts to disrupt and remove the threat of organized crime on society. The research conducted by Mohammad A. Tayebi and Uwe Glässer at Simon Fraser University provides an excellent stepping stone for intelligence and law enforcement agencies alike to more thoroughly analyze police/intelligence databases in ensuring the most useful allocation of policing resources

Director
Criminal Intelligence Services Ontario

Dr. Hugh Stevenson Ed.D.

Foreword

Preface

Predictive policing is promising for crime reduction and prevention to increase public safety, reduce crime costs to society, and protect the personal integrity and property of citizens. Strategic law enforcement operations aiming at proactive intervention in criminal activities can be a viable alternative to simply reacting to criminal acts. New methodologies in data science along with emerging applications of big data analytics to crime data promote a paradigm shift from tracking patterns of crime to predicting those patterns. Crime data analysis as presented in this book concentrates on relationships between offenders to better understand their criminal collaboration patterns through social network analysis. Law enforcement agencies have long realized the importance of co-offending networks for designing prevention and intervention strategies. According to Reiss (1988), understanding co-offending is central to understanding the etiology of crime and the effects of intervention strategies.

The objective of this book is to bring into focus predictive policing as a new paradigm in crime data mining and introduce social network analysis as a practical tool for turning crime data into actionable knowledge. The book systematically studies co-offending network analysis for various forms of criminal collaborations, starting with a formal model of crime data and co-offending networks to bridge the conceptual gap between abstract crime data and co-offending network mining. The formal representation of criminological concepts presented here allows computer scientists to think about algorithmic and computational solutions to problems long discussed in the criminology literature. This includes criminal network disruption, suspect investigation, organized crime group detection, co-offense prediction and crime location prediction. For each of the studied problems, we start with well-founded concepts and theories in criminology, then propose a computational model, and finally provide a thorough experimental evaluation, along with a discussion of the results. This way, the reader will be able to study the complete process of solving real-world multidisciplinary problems.

The targeted audience of this book includes researchers in computer science and criminology who are interested in predictive policing as an emerging multidisciplinary field as well as practitioners in collaborations between law enforcement and academia who search for novel and practical ideas to take predictive policing to the next level.

We would like to gratefully acknowledge the help and support of individuals and institutions who contributed to the work presented in this book, including RCMP "E" Division, BC Ministry for Public Safety and Solicitor General, Institute for Canadian Urban Research Studies (ICURS), Public Safety Canada, Patricia Brantingham, Paul Brantingham, Martin Ester, Gary Bass, Richard (Dick) Bent, Richard Frank, Mohsen Jamali, Vahid Dabbaghian, Laurens Bakker, and Austin Lawrence.

British Columbia, Canada

Mohammad A. Tayebi
Uwe Glässer

Contents

1 Introduction 1
References 5

2 Social Network Analysis in Predictive Policing 7
2.1 Conventional Crime Analysis 7
2.2 Predictive Policing 9
2.3 Social Network Analysis 9
2.4 Co-offending Networks 10
2.5 Co-offending Network Analysis in Practice 12
References 13

3 Structure of Co-offending Networks 15
3.1 Crime Data 15
3.1.1 Crime Data Model 16
3.1.2 Co-offending Network Model 17
3.1.3 BC Crime Dataset 18
3.2 Co-offending Network Structural Properties 19
3.2.1 Degree Distribution 20
3.2.2 Co-offending Strength Distribution 20
3.2.3 Connecting Paths 22
3.2.4 Clustering Coefficient 23
3.2.5 Connected Components Analysis 23
3.2.6 Network Evolution Analysis 25
3.3 Key Players in Co-offending Networks 28
3.3.1 Centrality Measures 28
3.3.2 Key Players Removal Effects 30
3.3.3 Experiments and Results 31
3.4 Conclusions 35
References 37

4 Organized Crime Group Detection 39
4.1 Background 40
4.1.1 Community Detection in Social Networks 42
4.2 Concepts and Definitions 44
4.2.1 Problem Definition 45
4.3 Proposed Approach 45
4.3.1 Organized Crime Group Detection 46
4.3.2 Organized Crime Group Evolution Model 48
4.4 Experiments and Results 49
4.4.1 Offender Groups Characteristics 49
4.4.2 Organized Crime Groups 55
4.5 Conclusions 61
References 62

5 Suspects Investigation 63
5.1 Background 64
5.2 Problem Definition 65
5.3 CrimeWalker 65
5.3.1 A Single Random Walk in CrimeWalker 66
5.3.2 CrimeWalker for a Set of Offenders 67
5.3.3 Similarity Measure for Offenders 68
5.3.4 Feature Weights Computation 69
5.4 Experiments and Results 69
5.4.1 Experimental Design 69
5.4.2 Comparison Partners 70
5.4.3 Experiments and Results 72
5.5 Conclusions 73
References 74

6 Co-offence Prediction 77
6.1 Background 79
6.1.1 Crime Prediction 79
6.1.2 Link Prediction 79
6.2 Concepts and Definitions 80
6.2.1 Notations 80
6.2.2 Offenders' Activity Space 81
6.2.3 Geographic and Network Proximity 81
6.2.4 Problem Definition 83
6.3 Supervised Learning for Co-Offence Prediction 83
6.3.1 Criminal Cooperation Opportunities 83
6.3.2 Reducing Class Imbalance Ratio 85
6.4 Prediction Features 87
6.4.1 Social Features 87
6.4.2 Geographic Features 87
6.4.3 Geo-Social Features 87
6.4.4 Similarity Features 89

6.5 Experiments and Results 89
6.5.1 Experimental Design 89
6.5.2 Single Features Significance 90
6.5.3 Prediction Evaluation 92
6.5.4 Criminological Implications 94
6.6 Conclusions 95
References 96

7 Personalized Crime Location Prediction 99
7.1 Background 101
7.1.1 Spatial Pattern of Crime 101
7.1.2 Crime Pattern Theory 102
7.1.3 Activity Space 102
7.1.4 Directionality 103
7.1.5 Crime Location Prediction 104
7.1.6 Urban Environment 105
7.1.7 Problem Definition 106
7.2 CRIMETRACER Model 106
7.2.1 Model Description 106
7.2.2 Random Walk Process 107
7.2.3 Starting Probabilities 109
7.2.4 Movement Directionality 110
7.2.5 Stopping Criteria 111
7.3 Experiments and Results 111
7.3.1 Data Characteristics 111
7.3.2 Experimental Design 113
7.3.3 Comparison Partners 114
7.3.4 Experiments and Results 116
7.4 Conclusions 124
References 124

8 Concluding Remarks 127
References 130

Index 131

Chapter 1
Introduction

Crime is a purposive deviant behavior that is an integrated result of different social, economical, and environmental factors [1]. Crime imposes a substantial cost on society at individual, community, and national levels [8]. Criminality worldwide makes trillions of dollars yearly, turning crime into one of the world's "top 20 economies" [5]. Based on the most recent report [6], the total cost of crime in Canada during 2012 is estimated as $81.5 billion, approximately 5.7 % of national income. Given such whopping costs, crime reduction and prevention strategies have become a top priority for law enforcement agencies.

Policymakers inevitably face enormous challenges deploying notoriously scarce resources even more efficiently to apprehend criminals, disrupt criminal networks, and effectively deter crime by investing in crime reduction and prevention strategies. While data collection from different sources, data preparation and information sharing pose difficult tasks, the big challenge for law enforcement agencies is analyzing and extracting knowledge from their large collection of crime data. Applying data-driven approaches on such data can provide a scientific foundation for developing effective crime reduction and prevention strategies through analysis of offenders' spatial decision making and their social standing. The main idea behind crime prediction techniques is that crime is not random but happens in patterned ways [2, 4, 9–13]. In the crime data mining process the goal is to understand criminal behaviors and extract criminal patterns in order to predict crime and take steps to prevent it.

Although crime analysis has a very long history, it has rapidly grown in the last decades to become common practice in law enforcement agencies. Crime analysis aims to assist police in criminal apprehension and crime reduction through systematic study of crime. Crime analysis has two main functions: strategic and tactical. Strategic analysis is about examining long-term crime trends. Tactical analysis concentrates on short-term and immediate problems to investigate the relationship between suspects and crime incidents.

M.A. Tayebi, U. Glässer, *Social Network Analysis in Predictive Policing*,
Lecture Notes in Social Networks, DOI 10.1007/978-3-319-41492-8_1

The rapid evolution of data science, employing techniques and theories drawn from broad areas such as machine learning and data mining, through availability of massive computational power increasingly influences our daily lives. Data are collected, modeled, and analyzed to uncover the patterns of human behavior and help with predicting social trends. This is changing the way we think about business, politics, education, health, and data science innovations will undoubtedly continue in the years to come. One particular area that has seen limited growth in accepting and using these powerful tools is public safety. This is somewhat surprising given the important role that predictive analytics can play in public safety.

New methodologies emerging in data science can advance crime analysis to the next level and move from tracking patterns of crime to predicting those patterns. This has led to a new paradigm of crime analysis, called *predictive policing*. Predictive policing uses data science to identify potential targets for criminal activity with the goal of crime prevention. Successful predictive policing results in more proactive policing and less reactive policing.

One of the most important goals of crime analysis is generating information that can enhance decision making for deploying police resources to prevent criminal activity. With predictive policing this process becomes more efficient and effective using the discovered patterns about crime locations, crime incidents, crime victims, criminals, criminal groups, and criminal networks. Nevertheless, predictive policing methods are neither a substitute for integrated solutions to policing nor equivalent to a crystal ball that can foretell the future. Predictive policing can facilitate proactive policing and improve intervention strategies by means of making efficient use of limited resources. These methods give law enforcement agencies a set of tools to do more with less.

One of the important tasks in predictive policing is analyzing the relationships between offenders to learn the criminal collaboration patterns. Law enforcement agencies have long realized the importance of analyzing co-offending networks—networks of offenders who have committed crimes together—for designing prevention and intervention strategies. Despite the importance of co-offending network analysis for public safety, computational methods for analyzing large-scale networks are rather premature.

Contrary to other social networks, concealment of activities and the identity of actors is a common characteristic of co-offending networks. Still, the network topology is a primary source of information for predictive tasks. Predictive policing methods can significantly take advantage of discovering collaboration patterns in co-offending networks. In this work we study co-offending network analysis as effective tool assisting predictive policing. The next section summarizes the contributions of this book.

This work is multidisciplinary, situated at the intersection of computer science and criminology, an area called *computational criminology* which uses computer science methods to formally define criminological problems, facilitate the process of understanding criminological phenomena, and present computational solutions for such problems. While computational modeling of crime can have far-reaching consequences on crime reduction and prevention, criminology and computer science still remain widely divided. This can be attributed to several factors such as the complicated nature of crime, challenges behind access to crime data, and lack of formal modeling of criminological issues. Formal modeling of a problem improves our understanding, and enhances formal analysis and reasoning. The initial problem formulation influences the rest of the research process. In multidisciplinary research problem formulation is a challenging task since it requires in-depth knowledge and good understanding from multiple domains.

The contribution of this work is two-fold. First, based on criminological theories, we formulate problems in the scope of predictive policing which can be addressed using social network analysis. It is important to point out the purpose of the work here is not alter or change the original problems, but present formal representations so that analysis can be done through algorithms. In the criminology literature there is a wide discussion on the problems studied here, but it lacks formal problem definitions required to make the problems tractable by computational models and methods. Our formal representation of criminological concepts allows computer scientists to think about algorithmic and computational solutions. Second, for each of the studied problems we propose a computational method, perform thorough experimental evaluation, and discuss the results.

We present here a unified crime data model as precise semantic foundation for co-offending network analysis [3]. This conceptual model provides a clear separation between crime data and computational methods, allowing the development of the computational methods to be done in a transparent way. We present a thorough study of structural properties of co-offending networks, and discuss implications of each of these properties for law enforcement agencies [3, 20]. Criminal network disruption strategies and verifying their impact on criminal groups is an important issue for police to control criminal groups. We study how centrality measures can be used to detect the key players in co-offending networks for the purpose of proactive interventions to control criminal organizations [17].

Organized crime is seen as a principal threat to public safety. Understanding organized crime as a multifaceted, dynamically changing form of criminality is very challenging. There have been some worthwhile studies [4], but there is no clear conceptualization of this phenomenon, and lack of clarity, transparency, and uncertainty creates obstacles to combat these organizations. While we are not aware of any formal modeling of organized crime groups in the literature, we present here a mathematical model of organized crime groups. From a social network analysis perspective we propose a community detection approach to identify organized crime groups, and a model to study their evolution trace [7, 14–16, 18].

We present a novel approach to crime suspect recommendation based on partial knowledge of offenders involved in a crime incident and a known co-offending network [19]. To solve this problem, we propose a random walk based method for recommending the top-N potential suspects.

The next problem we study is co-offence prediction. In the suspect investigation problem the goal is detecting potential suspects for a single crime incident, but in the co-offence prediction problem we aim at predicting the most probable criminal collaborations using the co-offending network structure and offenders' side information such as their demographic characteristics and spatial patterns. In the latter work, we propose a framework for co-offence prediction using supervised learning [22].

In our study of co-offence prediction, we realize the importance of the spatial movement patterns of offenders. After formalizing the concept of offenders' *probabilistic activity space*, as will be explained in Chap. 7, we propose an approach to generate the personalized activity space of an offender on a road network as urban layout. We use all available information about offenders in the crime dataset such as their crime records and co-offending network to enhance the method. Finally, we use the activity space of offenders to predict the location of their future crimes [21, 23, 24].

To the best of our knowledge, this work is the first comprehensive attempt to use co-offending network analysis in predictive policing suggesting a paradigm shift in the way co-offending network analysis is used for crime reduction and prevention. There are several major reasons that make this book a useful resource for readers with different backgrounds and goals: (1) We have explored thoroughly the criminology literature to identify and understand essential criminological problems that can take advantage of co-offending network analysis; therefore, this work covers the fundamental problems in this domain; (2) The proposed formal representation of the studied problems provides solid ground for algorithmic and computational research on those problems; (3) Our proposed algorithmic solutions for the studied problems have two important characteristics: first, they are established on the relevant criminological theories, and second, they are easy to interpret by domain experts including criminologists and law enforcement personnel; (4) The proposed methods are experimentally evaluated using a large real-world crime dataset producing high-quality results. We are not aware of any related work assessing performance using a similar dataset; and (5) This multidisciplinary work is completed in close collaboration with criminologists and law enforcement experts.

After this introductory chapter we provide an overview of co-offending network analysis applications in predictive policing in Chap. 2. We study general concepts of social network analysis and co-offending network analysis in this chapter. Chapter 3 discusses the structural properties of co-offending networks. This study helps to understand the basic properties of co-offending networks. The crime dataset used for experimental evaluation in this book is introduced in this chapter. In Chap. 4, we present our approach for detecting organized crime groups. Our proposed method for organized crime group detection is established on a comprehensive study of the concept of organized crime in the criminology literature, presented in the beginning of this chapter. Chapter 5 describes CRIMEWALKER, the proposed method for suspect investigation. We study how the structure of co-offending networks can be used in criminal profiling. In Chap. 6, we present a framework for co-offence prediction using supervised learning. More specifically, we study how different features of offenders can be used to predict a criminal collaboration. Chapter 7 describes CTIMETRACER, a method for personalized crime location prediction. CTIMETRACER generates the activity space of every offender for the purpose of predicting the location of their crimes. We study offender mobility to understand the activity space concept. Finally, we conclude this work and propose future work in Chap. 8. The chapters are self-contained with their own introduction, basic concepts, conclusions, and pointers to other relevant chapters or sections. They may be read in arbitrary order.

References

1. R. Boba, *Crime Analysis and Crime Mapping* (Sage, Thousand Oaks, 2013)
2. P.J. Brantingham, P.L. Brantingham, *Environmental Criminology* (Sage, Newbury Park, 1981)
3. P.L. Brantingham, M. Ester, R. Frank, U. Glässer, M.A. Tayebi, Co-offending network mining, in *Counterterrorism and Open Source Intelligence*, ed. by U.K. Wiil (Springer, Vienna, 2011), pp. 73–102
4. M. Carlo, *Inside Criminal Networks* (Springer, New York, 2009)
5. Crime one of world's 'top 20 economies' UN says (2012). Retrieved from http://www.cbc.ca/news/world/crime-one-of-world-s-top-20-economies-un-says-1.1186042
6. S. Easton, H. Furness, P. Brantingham, The cost of crime in canada (2014). Retrieved from www.fraserinstitute.org/uploadedFiles/fraser-ca/Content/research-news/research/publications/cost-of-crime-in-canada-2014.pdf
7. U. Glässer, M.A. Taybei, P.L. Brantingham, P.J. Brantingham, Estimating possible criminal organizations from co-offending data. *Public Safety Canada* (2012)
8. K.E. McCollister, M.T. French, H. Fang, The cost of crime to society: new crime-specific estimates for policy and program evaluation. Drug Alcohol Depend. **108**(1), 98–109 (2010)
9. J.M. McGloin, A.R. Piquero, On the relationship between co-offending network redundancy and offending versatility. J. Res. Crime Delinq. **47**(1), 63–90 (2009)
10. J.M. McGloin, C.J. Sullivan, A.R. Piquero, S. Bacon, Investigating the stability of co-offending and co-offenders among a sample of youthful offenders. Criminology **46**(1), 155–188 (2008)
11. A.J. Reiss Jr., Co-offending and criminal careers. Crime Justice **10**, 117–170 (1988)
12. D.K. Rossmo, *Geographic Profiling* (CRC Press, Boca Raton, 2000)
13. E.H. Sutherland, *Principles of Criminology* (J. B. Lippincott & Co., Chicago, 1947)
14. M.A. Tayebi, U. Glässer, Organized crime structures in co-offending networks, in *The 9th International Conference on Dependable, Autonomic and Secure Computing (DASC 2011)* (2011), pp. 846–853
15. M.A. Tayebi, U. Glässer, Crime group evolution in large co-offending networks, in *Proceedings of the 4th Annual Illicit Networks Workshop* (2012)
16. M.A. Tayebi, U. Glässer, Investigating organized crime groups: a social network analysis perspective, in *Proceedings of the 2012 International Conference on Advances in Social Networks Analysis and Mining (ASONAM'12)* (2012), pp. 565–572
17. M.A. Tayebi, L. Bakker, U. Glässer, V. Dabbaghian, Locating central actors in co-offending networks, in *Proceedings of the 2011 International Conference on Advances in Social Networks Analysis and Mining (ASONAM'11)* (2011), pp. 171–179
18. M.A. Tayebi, U. Glässer, P.L. Brantingham, Organized crime detection in co-offending networks, in *Proceedings of the 3rd Annual Illicit Networks Workshop* (2011)
19. M.A. Tayebi, M. Jamali, M. Ester, U. Glässer, R. Frank, CRIMEWALKER: a recommendation model for suspect investigation, in *Proceedings of the 5th ACM Conference on Recommender Systems (RecSys'11)* (2011), pp. 173–180
20. M.A. Tayebi, R. Frank, U. Glässer, Understanding the link between social and spatial distance in the crime world, in *Proceedings of the 20nd ACM SIGSPATIAL International Conference on Advances in Geographic Information Systems (ACM SIGSPATIAL GIS'12)* (2012), pp. 550–553

21. M.A. Tayebi, M. Ester, U. Glässer, P.L. Brantingham, CrimeTracer: activity space based crime location prediction, in *Proceedings of the 2014 International Conference on Advances in Social Networks Analysis and Mining (ASONAM'14)* (2014), pp. 472–480
22. M.A. Tayebi, M. Ester, U. Glässer, P.L. Brantingham, Spatially embedded co-offence prediction using supervised learning, in *Proceedings of the 20th ACM SIGKDD International Conference on Knowledge Discovery and Data Mining (KDD'14)* (2014), pp. 1789–1798
23. M.A. Tayebi, U. Glässer, P.L. Brantingham, Learning where to inspect: location learning for crime prediction, in *Proceedings of the 2015 International Conference on Intelligence and Security Informatics (ISI'15)* (2015), pp. 25–30
24. M.A. Tayebi, U. Glässer, M. Ester, P.L. Brantingham, Personalized crime location prediction. Eur. J. Appl. Math. **27**, 422–450 (2016)

Chapter 2
Social Network Analysis in Predictive Policing

Police departments have long used crime data analysis to assess the past, but the recent advances in the field of data science have introduced a new paradigm, called *predictive policing* which aims to predict the future. Predictive policing as a multidisciplinary approach brings together data mining and criminological theories which leads to crime reduction and prevention. Predictive policing is based on the idea that while some crime is random, the majority of it is not. In predictive policing crime patterns are learnt from historical data to predict future crimes.

Social connections and processes have a central role in criminology. But in the recent decades criminologists turned their attention to criminal networks to study the onset, maintenance, and desistance of criminal behavior [14]. More than two decades ago, Reiss [17] argued that "understanding co-offending is central to understanding the etiology of crime and the effects of intervention strategies." Meanwhile, influenced by increasing academic and societal awareness of the importance of social networks, law enforcement and intelligence agencies have come to realize the value of detailed knowledge of co-offending networks [4, 10, 14, 15, 17, 18].

In this chapter, we first discuss conventional crime analysis and predictive policing as a new perspective in crime-fighting strategies. Then, we introduce social network analysis and review general related work in co-offending network analysis. Finally, we briefly introduce different tasks of social network analysis in predictive policing studied in the next chapters of this book.

2.1 Conventional Crime Analysis

Analysis of crime has a long history, but *crime analysis* as a discipline is established when the first modern police started to work in London in the early nineteenth century [1]. After the constitution of the London police force in the 1820s, this force

M.A. Tayebi, U. Glässer, *Social Network Analysis in Predictive Policing*,
Lecture Notes in Social Networks, DOI 10.1007/978-3-319-41492-8_2

initiated a detective department with the responsibility of detecting crime patterns to solving crimes. The earliest source known for the term crime analysis is the book *police administration* published in 1963 [29]:

> The crime-analysis section studies daily reports of serious crimes in order to determine the location, time, special characteristics, similarities to other criminal attacks, and various significant facts that may help to identify either a criminal or the existence of a pattern of criminal activity. Such information is helpful in planning the operations of a division or district.

In the 1970s, the government of the USA tried to increase the ability of police departments in using crime analysis by inviting academics and practitioners. Later a group of academics started to emphasize the importance of characteristics of criminal events such as the location of crime which initiated the geographic analysis of crime. In the 1990s, with the increase of computer power, analyzing large crime dataset becomes computationally feasible, and police agencies tend to use crime analysis tools to generate analytical reports [19].

The main purpose of the crime analysis is crime reduction. In the policing approaches few mainstreams can be observed which get advantage of crime analysis [19]:

- **Standard model of policing.** The standard model of policing uses law enforcement in a reactive manner. Crime analysis helps in efficient allocation of police resources geographically and temporally.
- **Community policing.** Community policing strategies benefit from partnership and collaboration of the community to understand and solve the problems. The main role of crime analysis in these strategies is providing information to citizens.
- **Disorder policing.** Disorder policing or broken window policing is applying strict law enforcement procedures to minor offences to prevent happening of more serious crimes. Crime analysis is helpful in evaluating the disorder policing approaches.
- **Problem-oriented policing.** In problem-oriented policing the goal is diagnosing problems within the community and developing appropriate responses which solve the cause of the problems. Crime analysis is used in all phases of a problem-oriented policing strategy including scan, analysis, response, and assess.
- **Hotspots policing.** Hotspots policing is a location-based policing in which the police resources are allocated to different areas proportional to crime rate of each area. Crime analysis is used in identifying the hotspots.

Crime analysis contributed to the operational, tactical, and strategic police decision making for decades, but in the recent decade the emergence of data science field has arisen a new paradigm in this discipline called predictive policing introduced in the next section.

2.2 Predictive Policing

"Predictive policing refers to any policing strategy or tactic that develops and uses information and advanced analysis to inform forward-thinking crime prevention" [26], which involves multiple disciplines to form the rules and develop the models. Given that research strongly supports that crime is not random but rather occurs in patterns, the goal of predictive policing methods is to extract crime patterns from historical data at both macro and micro scales as a basis for prediction and prevention of future crimes [3, 8, 22–25]. This approach uses data-driven tools that benefit from data mining and machine learning techniques for predicting crime locations and temporal characteristics of criminal behavior.

Predictive analysis for policing can be divided into four classes:

- **Predicting offenders.** The goal is predicting future offenders using the history of individuals such as features of their living environment and behavioral patterns.
- **Predicting victims.** This is about identifying individuals who more likely than others may become victims and predicting risky situations for potential victims.
- **Predicting criminal collaborations.** Predicting likely future collaboration between offenders and the type of associated crime.
- **Predicting crime locations.** This task aims at predicting the location of future crimes at individual and aggregate level.

In this research our focus is on different problems related to the last two tasks: predicting criminal collaborations and crime locations. For solving this problems we use social network analysis methods. In the next sections we discuss social network analysis and its applications for predictive policing.

2.3 Social Network Analysis

Social networks represent relationships among social entities. Normally, such relationships can be represented as a network. Examples include interactions between members of a group (like family, friends, or neighbors) or economic relationships between businesses. Social networks are important in many respects. Social influence may motivate someone to buy a product, to commit a crime, and any other decision can be interpreted and modeled under a social network structure. Spread of diseases such as AIDS infection and the diffusion of information and word of mouth also strongly depend on the topology of social networks.

Social network analysis (SNA) focuses on structural aspects of networks to detect and interpret the patterns of social entities [28]. SNA essentially takes a network with nodes and edges and finds distinguished properties of the network through formal analysis. Data mining is the process of finding patterns and knowledge hidden in large databases [9]. Data mining methods are increasingly being applied to social networks, and there is substantial overlap and synergy with SNA.

New techniques for the analysis and mining of social networks are developed for a broad range of domains, including health [27] and criminology [31]. These methods can be categorized depending on the level of granularity at which the network is analyzed [2]: (1) methods that determine properties of the social network as a whole; (2) methods that discover important subnetworks; (3) methods that analyze individual network nodes; and (4) methods that characterize network evolution. In the following, we list the primary tasks of SNA:

- *Centrality analysis* [28] aims at determining more important actors of a social network so as to understand their prestige, importance, or influence in a network.
- *Community detection* [6] methods identify groups of actors that are more densely connected among each other than with the rest of the network.
- *Information diffusion* [12] studies the flow of information through networks and proposes abstract models of that diffusion such as the Independent Cascade model.
- *Link prediction* [13] aims at predicting for a given social network how its structure evolves over time, that is, what new links will likely form.
- *Generative models* [5] are probabilistic models which simulate the topology, temporal dynamics, and patterns of large real-world networks.

SNA also greatly benefits from visual analysis techniques. Visualizing structural information in social networks enables SNA experts to intuitively make conclusions about social networks that might remain hidden even after getting SNA results. Different methods of visualizing the information in a social network providing examples of the ways in which spatial position, color, size, and shape can be used to represent information are mentioned in [7].

In the next section we introduce co-offending networks as a special type of social networks.

2.4 Co-offending Networks

Criminal organizational systems differ in terms of their scope, form, and content. They can be a simple co-offending looking for opportunistic crimes, or a complex organized crime group involved in serious crimes. They can be formed based on one-time partisanship for committing a crime, or their existence can have continuity over time and across different crime types [4]. In a criminal organization system interaction among actors can be initiated from family, friendship, or ethnic ties. Here, our focus is on co-offending networks.

A *co-offending network* is a network of offenders who have committed crimes together [17]. With increasing attention to SNA, law enforcement and intelligence agencies have come to realize the importance of detailed knowledge about co-offending networks. Groups and organizations that engage in conspiracies, terroristic activities and crimes like drug trafficking typically do this in a concealed fashion, trying to hide their illegal activities. In analyzing such activities, investigations do not only focus on individual suspects but also examine criminal groups and illegal organization and their behavior.

Thus, it is important to identify co-offending networks in data resources readily available to investigators, such as police arrest data and court data, and study them using social network analysis methods. In turn, social network analysis can provide useful information about individuals as well. For example, investigators could determine who are key players, and subject them to closer inspection. In general, knowledge about co-offending network structures provides a basis for law enforcement agencies to make strategic or tactical decisions.

Several empirical studies that use social network analysis methods to analyze co-offending networks have focused on the stability of associations in such networks. Reiss [17] concludes that the majority of co-offending groups are unstable, and their relationships are short-lived. This is corroborated by McGloin et al. [15], who showed that there is some stability in co-offending relationships over time for frequent offenders, but in general, delinquents do not tend to reuse co-offenders. Reiss et al. [18] also found that co-offenders have many different partners, and are unlikely to commit crimes with the same individuals over time. However, Reiss [17] also states that high frequency offenders are "active recruiters to delinquent groups and can be important targets for law enforcement." It should be noted that the findings of these works were obtained on very small datasets: 205 individuals in [18], and 5600 individuals in [15], and may therefore not be representative.

These studies only analyzed co-offending networks. Smith [21] widened the scope of co-offending network analysis, enhancing the network by including extra information, particularly for the purpose of criminal intelligence analysis. For example, nodes of the network could be offenders, but also police officers, reports, or anything that can be represented as an entity. Links are associated with labels which denote the type of the relationship between the two entities, such as "mentions" or "reported by." A similar approach was taken by Kaza et al. [11], who explored the use of criminal activity networks to analyze information from law enforcement and other sources for transportation and border security. The authors defined the criminal activity network as a network of interconnected criminals, vehicles, and locations based on law enforcement records, and concluded that including especially vehicular data in criminal activity network is important, because vehicles provide new investigative points.

A slightly different take on widening the scope of co-offending network analysis was taken by Xu et al. [30], who employed the idea of a "concept space" in order to establish the strength of links between offenders. Not only the frequency of co-offending, but also event and narrative data were used to construct an undirected but weighted co-offending network. The goal was to identify central members and communities within the network, as well as interactions between communities. By applying cluster analysis in order to detect subgroups within the network they were able to detect overall network structures which could then be used by criminal investigators to further their investigations.

COPLINK [10] was one of the first large-scale research projects in crime data mining, and an excellent work in criminal network analysis. It is remarkable in its practicality, being integrated with and used in the workflow of the Tucson Police Department. Xu et al. [31] built on this when they created CrimeNet Explorer, a framework for criminal network knowledge discovery incorporating hierarchical clustering, SNA methods, and multidimensional scaling. The authors further expanded the research in [30] and designed a full-fledged system capable of incorporating external data, such as phone records and report narratives, in order to establish stronger ties between individual offenders. Their results were compared to the domain knowledge offered by the Tucson Police Department, whose jurisdiction the data came from.

2.5 Co-offending Network Analysis in Practice

Co-offending network analysis contributes to predictive policing by detecting hidden links and predicting potential links among offenders. In this section, we introduce important applications of co-offending network analysis in predictive policing which are covered in this research.

- **Co-offending network disruption.** Actors of a social network can be categorized based on their relations in the network. Actors in the same category may take similar roles within an organization, community, or whole network. These roles are usually depend on the network structure and the actors' position in the network. For instance, actors who are located in the central positions of a social network may be detected as key players in that network. Actors who are connected to many other actors may be viewed as socially active players, and actors who are frequently observed by other actors may be identified as popular players.

  In the co-offending networks disruption problem the goal is finding a set of players whose removal creates a network with the least possible cohesion. In other words, their removal maximally destabilizes the network. This task is critical in the co-offending network analysis where removing the key players may sabotage the network and decrease the aggregate crime rate. We study this problem in Chap. 3.
- **Organized crime group detection.** Organized crime is a major international concern. Organized crime groups produce disproportionate harm to societies, and an increasing volume of violence is related to their activities. Since the aim of organized crime groups is gaining material benefit they try to access to resources that can be profitably exploited. In terms of economic-related crimes (e.g., credit and debit card fraud) organized crime costs Canadians five billion dollar a year [20].

  Understanding the structure of organized crime groups and the factors that impact on it is crucial to combat organized crime. There are several possible perspectives how to define the structure of organized crime groups, but recent criminological studies are increasingly focusing on using social network analysis for this purpose. The idea of using social network analysis is that links between offenders and subgroups of an organized crime group are critical determinant of the performance and sustainability of organized crime groups [16]. In Chap. 4, we study the organized crime group detection problem.

- **Suspect investigation.** Security services can more precisely focus their efforts based on probable relationships in criminal networks that have previously not observed. Traditional suspect investigation methods use partial knowledge discovered from crime scene to identify potential suspects. Co-offending network analysis as a complement of criminal profiling methods can contribute to the suspect investigation task in cases with multiple offenders committing a crime, but a subset of offenders are charged. This issue is addressed in Chap. 5.
- **Co-offence prediction.** Link prediction is an important task in social network analysis that can help to study and understand the network structure. Link prediction methods can be used to extract missing information, identify hidden links, evaluate network evolution mechanisms, and so on. Co-offence prediction can be defined as link prediction problem for co-offending networks. Chapter 6 is about the co-offence prediction problem.
- **Personalized crime location prediction.** An important aspect of crime is the geographic location that crime happens. Every neighborhood provides some condition in which criminal behavior takes place, but crime distribution in city neighborhoods is not even. Understanding the spatial patterns of crime is essential for law enforcement agencies to design efficient crime reduction and prevention policies. Although mining spatial patterns of crime data in the aggregate level took special attention in the criminology literature, there is not that much work about crime spatial patterns for individual offenders. This problem is addressed in Chap. 7.

References

1. R. Boba, *Crime Analysis and Crime Mapping* (Sage, Thousand Oaks, 2013)
2. U. Brandes, T. Erlebach, *Network Analysis: Methodological Foundations*. Lecture Notes in Computer Science/Theoretical Computer Science and General Issues (Springer, Heidelberg, 2005)
3. P.L. Brantingham, M. Ester, R. Frank, U. Glässer, M.A. Tayebi, Co-offending network mining, in *Counterterrorism and Open Source Intelligence*, ed. by U.K. Wiil (Springer, Vienna, 2011), pp. 73–102
4. M. Carlo, *Inside Criminal Networks* (Springer, New York, 2009)
5. D. Chakrabarti, C. Faloutsos, Graph mining: laws, generators, and algorithms. ACM Comput. Surv. **38**(1), Article 2 (2006)
6. J. Chen, O.R. Zaïane, R. Goebel, Detecting communities in social networks using max-min modularity, in *Proceedings of SIAM International Conference on Data Mining (SDM'09)* (2009), pp. 978–989
7. L.C. Freeman. Visualizing social networks. J. Soc. Struct. **1**(1), 4 (2000)
8. U. Glässer, M.A. Taybei, P.L. Brantingham, P.J. Brantingham, Estimating possible criminal organizations from co-offending data. *Public Safety Canada* (2012)
9. J. Han, M. Kamber, *Data Mining: Concepts and Techniques* (Morgan Kaufmann, San Francisco, 2006)

10. R.V. Hauck, H. Atabakhsh, P. Ongvasith, H. Gupta, H. Chen, Using Coplink to analyze criminal-justice data. Computer **35**(3), 30–37 (2002)
11. S. Kaza, H. Chen, Effect of inventor status on intra-organizational innovation evolution, in *Proceedings of the 42nd Hawaii International Conference on System Sciences (HICSS'09)* (2009), pp. 1–10
12. D. Kempe, J. Kleinberg, É. Tardos, Influential nodes in a diffusion model for social networks. Autom. Lang. Program. **3580**, 1127–1138 (2005)
13. D. Liben-Nowell, J. Kleinberg, The link prediction problem for social networks, in *Proceedings of the 12st ACM international conference on Information and knowledge management (CIKM'03)* (2003), pp. 556–559
14. J.M. McGloin, A.R. Piquero, On the relationship between co-offending network redundancy and offending versatility. J. Res. Crime Delinq. **47**(1), 63–90 (2009)
15. J.M. McGloin, C.J. Sullivan, A.R. Piquero, S. Bacon, Investigating the stability of co-offending and co-offenders among a sample of youthful offenders. Criminology **46**(1), 155–188 (2008)
16. C. Moselli, T. Gabor, J. Kiedrowski, The factors that shape organized crime. *Public Safety Canada* (2010)
17. A.J. Reiss Jr., Co-offending and criminal careers. Crime Justice **10**, 117–170 (1988)
18. A.J. Reiss Jr., D.P. Farrington, Advancing knowledge about co-offending: results from a prospective longitudinal survey of london males. J. Crim. Law Criminol. **82**, 360–395 (1991)
19. R.B. Santos, *Crime Cnalysis with Crime Mapping* (Sage, Thousand Oaks, 2012)
20. Serious and organized crime (2015). Retrieved from http://www.rcmp-grc.gc.ca/soc-cgco/index-eng.htm
21. M.N. Smith, P.J.H. King, Incrementally visualising criminal networks, in *Proceedings of the Sixth International Conference on Information Visualisation (IV'02)* (2002), pp. 76–81
22. M.A. Tayebi, U. Glässer, Investigating organized crime groups: a social network analysis perspective, in *Proceedings of the 2012 International Conference on Advances in Social Networks Analysis and Mining (ASONAM'12)* (2012), pp. 565–572
23. M.A. Tayebi, L. Bakker, U. Glässer, V. Dabbaghian, Locating central actors in co-offending networks, in *Proceedings of the 2011 International Conference on Advances in Social Networks Analysis and Mining (ASONAM'11)* (2011), pp. 171–179
24. M.A. Tayebi, R. Frank, U. Glässer, Understanding the link between social and spatial distance in the crime world, in *Proceedings of the 20nd ACM SIGSPATIAL International Conference on Advances in Geographic Information Systems (ACM SIGSPATIAL GIS'12)* (2012), pp. 550–553
25. M.A. Tayebi, M. Ester, U. Glässer, P.L. Brantingham, Spatially embedded co-offence prediction using supervised learning, in *Proceedings of the 20th ACM SIGKDD International Conference on Knowledge Discovery and Data Mining (KDD'14)* (2014), pp. 1789–1798
26. C. Uchida, *A National Discussion on Predictive Policing: Defining Our Terms and Mapping Successful Implementation Strategies* (National Institute of Justice, Washington, 2012)
27. T.W. Valente, *Social Networks and Health: Models, Methods, and Applications* (Oxford University Press, Oxford, 2010)
28. S. Wasserman, K. Faust, *Social Network Analysis: Methods and Applications* (Cambridge University Press, New York, 1994)
29. O.W. Wilson, *Police Administration* (McGraw-Hill, New York, 1963)
30. J.J. Xu, H. Chen, Untangling criminal networks: a case study. Intell. Secur. Inform. **2665**, 958–958 (2003)
31. J.J. Xu, H. Chen, CrimeNet explorer: a framework for criminal network knowledge discovery. ACM Trans. Inf. Syst. **23**(2), 201–226 (2005)

Chapter 3
Structure of Co-offending Networks

Co-offending networks are generally extracted from police recorded crime data. For doing so, we need to have a clear view of crime data. In this chapter, we first introduce a unified formal model of crime data as a semantic framework for defining in an unambiguous way the meaning of co-offending networks at an abstract level. Then, we introduce a real-world crime dataset, referred to as *BC crime dataset* which is used in this book, and the *BC co-offending network* which is extracted from this dataset. The BC crime dataset represents 5 years of police arrest-data for the regions of the Province of British Columbia which are policed by the RCMP, comprising several million data records.

The structure of social networks affects the process of human interaction and communication such as information diffusion and opinion formation. Studying structural properties of a social network is essential for understanding the social network. The same statement is true about co-offending networks. In the second part of this chapter, we study structural properties of the BC co-offending network and discuss important implications of such properties for law enforcement agencies. In the last part of this chapter, we focus on detecting key players of co-offending networks, and how this aspect contributes to co-offending network disruption.

Section 3.1 introduces the crime data. Section 3.2 presents structural properties of co-offending networks. We study how to identify key players of a co-offending network in Sect. 3.3. Section 3.4 concludes this chapter.

3.1 Crime Data

Police recorded crime data is highly sensitive making it difficult for the researchers to access in a convenient way. Researchers obtain access to crime data if only they provide high standards of safe data storage and processing solutions. Some of the

M.A. Tayebi, U. Glässer, *Social Network Analysis in Predictive Policing*,
Lecture Notes in Social Networks, DOI 10.1007/978-3-319-41492-8_3

preconditions for a researcher to access crime data include signing confidentiality agreements, maintaining comprehensive security measures for crime data storage and retrieval, and finally providing police background checks.

This section proposes a unified formal model of crime data serving as the semantic framework for defining in a concise and unambiguous way properties of interest in the analysis of co-offending networks and their constituent entities. Specifically, the formal model aims at bridging the conceptual gap between data level, mining level and interpretation level, and facilitates separating the description of data from the details of data mining and analysis. By gradually transforming and reducing the unified model to more specific views, the co-offending network model is obtained as one such view.

3.1.1 Crime Data Model

We model a crime dataset based on a collection of regular police records that document crime events in a geographic area of interest reported over some period of time [5]. Each record refers to a single crime event; two or more records may refer to the same event. A crime dataset abstractly represents a finite set of crime events as associated with a given collection of regular police records such that each single event, together with all reported data and information related to this event, is identified with a different element in the crime data set and every element in this set uniquely refers to one of the crime events.

Formally, we represent the logical organization of the crime data and information associated with a crime dataset $\mathscr{C}$ as a finite graph structure in the form of an attributed tripartite *hypergraph* $\mathscr{H}(\mathscr{N},\mathscr{E})$ with a set of nodes $\mathscr{N}$ and a set of hyperedges $\mathscr{E}$. The set $\mathscr{N}$ is composed of three disjoint subsets, $A = \{a_1, a_2, \ldots, a_q\}$, $I = \{i_1, i_2, \ldots, i_r\}$, and $R = \{r_1, r_2, \ldots, r_s\}$, respectively, representing *actors* like offenders, suspects, victims, witnesses, and bystanders; *incidents* referring to reported crime events; and *resources* used in a crime, such as weapons, tools, mobile phones, vehicles, and bank accounts. Generic actors serve as placeholders if a person's identity remains unclear, say an unrecognized offender who evaded apprehension. Whenever no specific resource can be identified or has been reported, the distinguished element "*unknown*" is used as a placeholder.

A hyperedge e of $\mathscr{E}$ is a non-empty subset of nodes $\{n_1, n_2, \ldots n_p\} \subseteq \mathscr{N}$ such that the following three conditions hold: $|e \cap I| = 1$, $|e \cap A| \geq 1$ and $|e \cap R| \geq 1$. For any $e, e' \in \mathscr{E}$ with $e \cap I = e' \cap I$, it follows that $e = e'$. Intuitively, a hyperedge e of $\mathscr{H}$ associates a set of actors $\{a_{i_1}, a_{i_2}, \ldots, a_{i_j}\} \subseteq A$ and a set of resources $\{r_{i_1}, r_{i_2}, \ldots, r_{i_l}\} \subseteq R$ with a crime incident $i_k \in I$, where $e = \{i_k, a_{i_1}, a_{i_2}, \ldots, a_{i_j}, r_{i_1}, r_{i_2}, \ldots, r_{i_l}\}$ as illustrated in Fig. 3.1.

Finally, with each node $n \in \mathscr{N}$ we associate some non-empty list of attributes characterizing the entity represented by n. Attributes of actors, for instance include the name, address and contact details, and the criminal profile information of known offenders while attributes of incidents include the crime type, the time of the

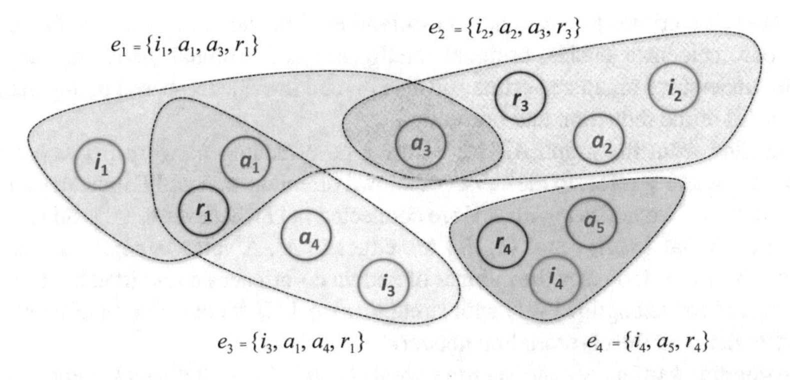

Fig. 3.1 Hypergraph $\mathscr{H}$ (without attributes) for a simple crime data model $\mathscr{C}$

incident, longitude and latitude coordinates of the crime location, and the role of each person identified in connection with the incident, among various other types of data and information.

For analyzing and reasoning about specific aspects of crime data that can be described in terms of entities and their relations, the hypergraph $\mathscr{H}$ is transformed in several steps into several bipartite graph structures as follows. From the original graph $\mathscr{H}$, we derive a hypergraph $\mathscr{H}'(\mathscr{N}, \mathscr{E}')$, where $\mathscr{N}$ is identical to the node set of $\mathscr{H}$ and

$$\mathscr{E}' = \{\{a,i,r\} \mid \exists e \in \mathscr{E} : \{a,i,r\} \subseteq e,\ a \in A, i \in I, r \in R\}.$$

Note that $\mathscr{H}'$ has the same attributes as $\mathscr{H}$. Now, $\mathscr{H}'$ can further be decomposed in a straightforward way into three *bipartite* graphs that, respectively, model the relations between actors and incidents (graph *AI*), actors and resources (graph *AR*), and incidents and resources (graph *IR*). The goal of the transformation process is focusing on more important subsets of data to extract more meaningful elements from the crime dataset. This multi-step process not only facilitates the extraction of more important and meaningful elements of crime data, but also it gives us a better understanding of different aspects of crime data. From each of these bipartite graphs we can extract a set of networks and use them for different mining purposes. For instance, we can extract co-offending network or two-mode network of offenders and victims from the graphs *AI* to learn the patterns among offenders, and the patterns between offenders and victims.

3.1.2 Co-offending Network Model

A co-offending network consists of groups of two or more offenders who have committed crimes together. Co-offending networks constitute a widespread form of social networks that is of considerable interest in crime investigations and in the study of crime. For instance, co-offending behavior is a relevant factor for law enforcement agencies, criminal intelligence and criminal justice agencies to better understand organized crime and also pivotal in evidence-based policy making aiming at crime reduction and prevention.

Starting from the graph AI, we define a co-offending network [5] as a graph $G(V,E)$, where V refers to the subset of known offenders in A and E indicates known co-offences. Two nodes $a_m, a_n \in V$ are connected in G whenever there is an incident $i_k \in I$ such that $\{a_m, i_k\}$ and $\{a_n, i_k\}$ are edges in AI. A value *strength* assigned to each edge e in E indicates the number of known co-offences committed by the same two offenders, $\text{strength}(e) \in \mathbb{N}$ with $\text{strength}(e) \geq 1$. Γ_i denotes the set of neighbors of offender a_i in the co-offending network.

Assuming k offenders and m crime events $(k, m > 1)$, we define a $k \times m$ matrix M such that $m_{uv} = 1$, if offender u is involved in event v, and "0" otherwise. This way, we can express the co-offending network as a $k \times k$ matrix $N = MM^T$ and therefore have

$$n_{u,v} = \sum_{x=1}^{k} n_{ux} n_{xv} \tag{3.1}$$

This matrix links offenders involved in the same crime events. For any two given offenders, the strength of a link is the number of co-offenses. The diagonal of this matrix shows for each offender the number of related crime events.

3.1.3 BC Crime Dataset

As a result of a research memorandum of understanding between ICURS[1] and "E" Division of Royal Canadian Mounted Police (RCMP) and the Ministry of Public Safety and the Solicitor General, 5 years of real-world crime data was made available for research purposes. This data was retrieved from the RCMP's Police Information Retrieval System (PIRS), a large database system keeping information for the regions of the Province of British Columbia which are policed by the RCMP. PIRS contains information about all reported crime events (≈4.4 million) and all persons (offenders, victims, witnesses, etc.) associated with a crime incident (≈9 million referring to about four million unique individuals). Table 3.1 shows the statistical properties of the BC co-offending network.

In total, there are 39 different subject (person) groups. For any given crime incident, every related subject has up to three different status fields, stating the subject's "role" in this incident. Out of four million subjects in the dataset, 250,492, 255,302, 190,406, and 228,792, respectively, appear at least once as charged,

[1]The Institute for Canadian Urban Research Studies (ICURS) is a university research center at Simon Fraser University.

Table 3.1 Statistical properties of the BC co-offending network

Metric	All crimes	Serious	Property	Drugs	Moral
Number of offenders	157,274	31,132	44,321	54,286	35,266
Average degree	4	1.85	1.95	2.15	4.8
Average distance	12.2	1.69	8.45	22.17	3.41
Diameter	36	13	24	56	19
Effective diameter	16.87	4.1	14.36	36.14	5.68
Clustering coefficient	0.39	0.28	0.33	0.39	0.49
Largest component percentage	25 %	10 %	32 %	23 %	21 %

chargeable, charge recommended, or suspect. In our experiments, we restrict on the subjects in these four categories. Being in one of these categories means that the police were serious about the subjects involvement in a crime. In this book, we call this group of subjects "offenders."

In total, there are over 50 groups of crime types. Four most important groups are:

- *Serious Crimes*: crimes against a person, such as homicide and attempted homicide, assault, abduction;
- *Property Crimes*: crimes against property, such as burglary (break and enter into a premises or real property, and theft);
- *Moral Crimes*: such as prostitution, arson, child pornography, gaming, breach;
- *Drug Crimes*: such as trafficking, possession, import/export.

3.2 Co-offending Network Structural Properties

In this section, we present the important concepts of structural analysis of social networks as well as the results of our analysis on the BC co-offending networks [5]. We apply the analysis tasks on the co-offending networks extracted from different crime types and also on several snapshots of these networks.[2] $G_u(t)$ denotes the co-offending network of a specific crime type u (a, s, p, d, and m represent the all, serious, property, drugs, and moral crimes types) from year 2001 to year t.

[2]In implementing the analysis tasks, we used SNAP library which is publicly available at http://snap.stanford.edu/.

3.2.1 Degree Distribution

The degree of a node is the number of edges the node has. The *degree distribution*, $P(k)$, gives the probability that a randomly selected node has k links. Studies have shown that the most real-world networks from diverse fields ranging from sociology to biology to communication follow a *power-law distribution* [1]:

$$P(k) = k^{-\lambda} \tag{3.2}$$

where λ is called *exponent* of the distribution. Power-law distribution implies that nodes with few links are numerous, while very few nodes have very large number of links. Networks with this property are called *scale free networks*.

There are other network models such as Erdos–Renyi [11] and Watts and Strogatz [1] models that are known as exponential networks, and their degree distribution conforms to a Poisson distribution. In this type of networks there is a peak at the average degree of network, therefore, most of the nodes have the same degree around average degree of network and very few nodes have very small or very large node degrees.

Our studied co-offending network is scale free. Figure 3.2 demonstrates the cumulative degree distribution for different types of co-offending networks. Degree distribution of all of these networks are consistent with the power-law distribution. Meaning that the majority of offenders have small degree, and a few offenders have significantly higher degree. To test how well the degree distributions are modeled by a power-law, we computed the best power-law fit using the maximum likelihood method [9]. The power-law exponent for all crimes, serious, property, drugs and moral co-offending networks, respectively, are 2.29, 1.57, 1.42, 1.53, and 2.28.

3.2.2 Co-offending Strength Distribution

Each link in a co-offending network is associated with a co-offending strength. The co-offending strength of two offenders i and j is equal to the number of crimes these offenders committed together. We define network $\bar{G}(V,E,\alpha)$ where E includes the links between the pairs of offenders $i,j \in V$ whose co-offending strengths exceed a specified threshold α. Then we will have a family of networks $\{\bar{G}(\alpha_1), \bar{G}(\alpha_2), \ldots, \bar{G}(\alpha_m)\}$ corresponding to different values of α.

Figure 3.3 plots the distribution of number of nodes and links for the threshold networks. Again, a power-law distribution of co-offending strength suggests that the vast majority of dyads offended once or twice, but there are only about hundred dyads that offended with each other more than ten times over 5 years. When two offenders collaborate on multiple incidents, the likelihood of having a strong relationship between them is higher. Therefore, such offenders and their behaviors should be inspected more carefully by the crime investigators.

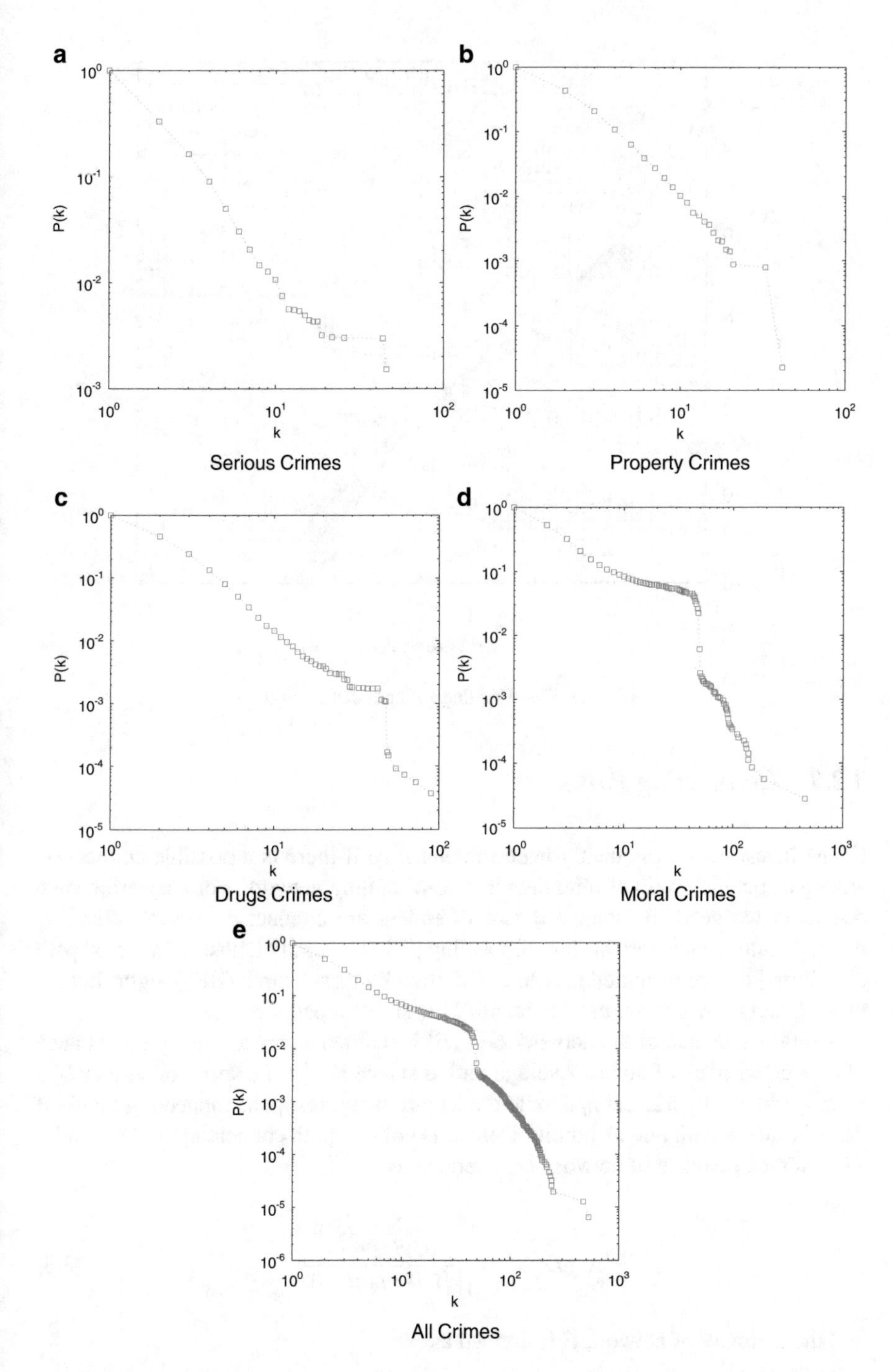

Fig. 3.2 Degree distribution of co-offending networks for different crime types. (**a**) Serious crimes. (**b**) Property crimes. (**c**) Drug crimes. (**d**) Moral crimes. (**e**) All crimes

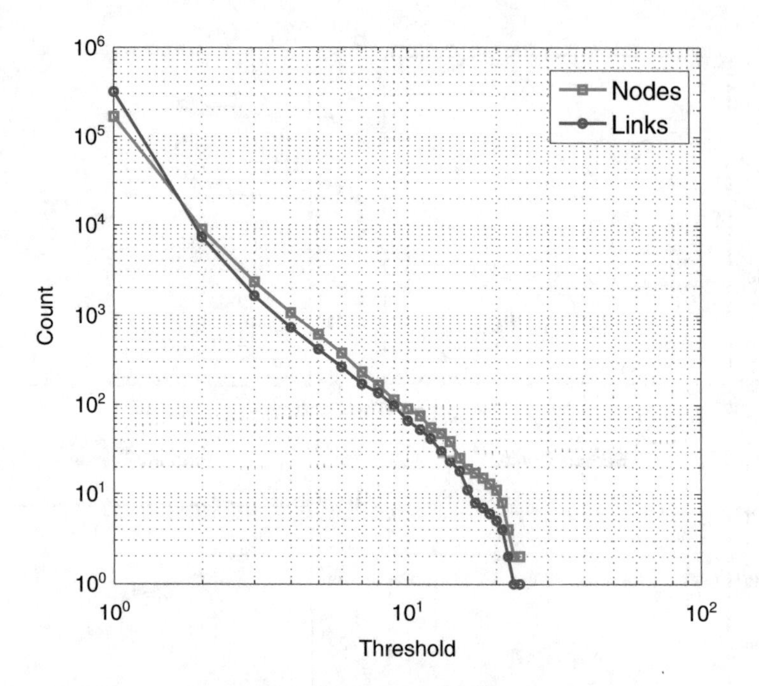

Fig. 3.3 Co-offending strength distribution

3.2.3 Connecting Paths

Crime investigators frequently need to determine if there is a possible connection among a specific group of offenders in a co-offending network. For answering such questions we need to identify if two offenders are connected in a co-offending network, and what is the shortest connecting path. Generally Dijkstra's shortest path algorithm [10] for weighted networks and Breadth First Search (BFS) algorithm for unweighted networks are used to identify the shortest paths.

Average distance of the network $G(V,E)$ is defined as the average path distance of connected pairs of nodes. Average path distance shows the speed of spreading a message in a network. Let l_{ij} denote the length of shortest path connecting i and j if there is such a path and as infinity if there is not any path connecting nodes i and j. The average distance of network G is defined as

$$\text{Avg}D(G) = \frac{\sum_{\{i,j\}:l_{ij}\neq\infty} l_{ij}}{|\{\{i,j\} : l_{ij} \neq \infty\}|} \tag{3.3}$$

and the diameter of network G is defined as

$$\text{Diam}(G) = \text{Max}(l_{ij} : l_{ij} \neq \infty) \tag{3.4}$$

Diameter is the length of the longest shortest path between any pair of nodes, which describes the compactness and connectivity of the network. A network with a small diameter is well connected but a network with a large diameter is sparsely connected. For removing the effect of outliers another measure called effective diameter is used. Effective diameter is the minimum number of hops in which at least 90 % of all connected pairs of nodes can reach each other [17]. Table 3.1 shows the average distance, diameter and effective diameter for the five studied co-offending networks. The average distance and diameters for some of them are remarkably short. For instance, for the network $G_a(2006)$ average distance, diameter, and effective diameter are 12.2, 36, and 16.87, respectively.

3.2.4 Clustering Coefficient

In many social networks friends of an actor is likely to be also her friend. In other words, actors tempt to create complete triangles of relationships. This property is called network clustering or transitivity. The clustering coefficient of a node in a co-offending network tells us how much an offender's collaborators are willing to collaborate with each other. Local clustering coefficient calculates the probability of neighbors of a node to be neighbors to each other is given by

$$C_v = \frac{a_v}{|\Gamma_v|(|\Gamma_v|-1)} \quad (3.5)$$

where $|\Gamma_v|$ is the number of neighbors of v. $|\Gamma_v|(|\Gamma_v|-1)$ is the maximum number of links that can exist between neighbors of v, and a_v is the number of links that actually exist among neighbors of v. The clustering coefficient of the network is computed by averaging C_v over all nodes [1]:

$$C = \frac{1}{|V|}\sum_{v \in V} C_v \quad (3.6)$$

The clustering coefficient of $G_s(2006)$, $G_p(2006)$, $G_d(2006)$, $G_m(2006)$, and $G_a(2006)$ are, respectively, 0.28, 0.33, 0.39, 0.49, and 0.39. The clustering coefficient of a network shows to what extent friends of a person are also friends with each other. Co-offending network of moral crimes has larger clustering coefficient, which shows offenders in this network have closer collaboration comparing to other types of co-offending networks.

3.2.5 Connected Components Analysis

Entities of a network are interested in forming groups and interact more closely to each other inside the group. The specific characteristic of a group is that there is a higher degree of connectivity inside the group than entities outside the group. Nowadays, studying the behavior of criminal groups becomes more important. In the last decade there have been more and more experimental studies into criminal activities that need specific forms of collaboration and organization [7]. For detecting these types of collaboration we need to mathematically formalize concepts such as offender group, gang, organized crime, and corporate crime and then design efficient algorithms for this purpose. By inspecting relations between offenders to identify criminal groups, law enforcement organizations can track the origin and core of what may become an organized crime group or a gang. In this way a criminal group can be identified prior to its formation and police can follow such offenders' behavior.

As a first step, we studied the distribution of size of *connected components* in the co-offending networks. A connected component is a subset of network where there exists a path between any two nodes in it [22]. If two offenders were involved in a crime, there is a path between them. If a third offender commits a crime with one of these offenders, a path can be built connecting the first offender with the third offender and so on. If a path between two offenders can be established, the two offenders belong to the same component. Studying characteristics of connected components is an initial step in analysis of epidemic spreading through a social network. Extracting patterns of connected components of co-offending networks provides valuable information for law enforcement agencies in fighting epidemics of crime.

Let $|c|$ represent the size of component c. Then we define three types of components: large components $|c| \geq 1000$, medium components $100 \leq |c| < 1000$, and small sized components $2 \leq |c| \leq 100$. In the network $G_a(2006)$, 25 %, 1 %, and 74 % of the whole offenders are connected to each other, respectively, through large, medium, and small components.

In the second step, we study the community structure in the co-offending network. To do this, we apply the Girvan–Newman algorithm [12] for detecting communities on the network $G_a(2006)$. The key idea behind this algorithm is that the edges connecting highly clustered communities have a higher edge betweenness, and communities can be detected by progressively removing such edges from the network. After every removal, the betweenness of edges is recalculated, and the process is repeated until the network is divided into a specified number of subnetworks, the communities.

Figure 3.4 shows the size distribution of detected communities and connected components. The largest extracted community size has about 4000 nodes, which is relatively small compared to the largest component with more than 39,000 nodes. However, a criminal group of few thousand members cannot be interpreted from a criminological perspective. There is a need for novel community extraction methods that particularly address the special requirements of co-offending networks. In Chap. 4, we study the problem of detecting organized crime groups.

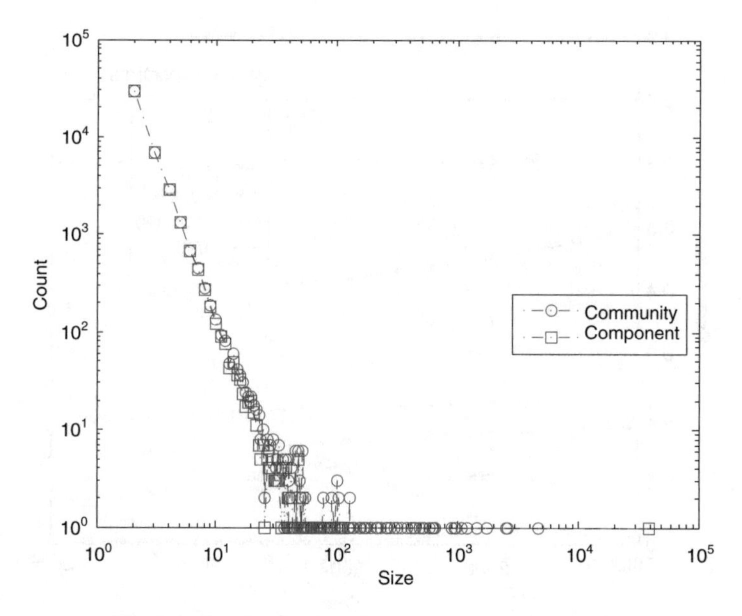

Fig. 3.4 Size distribution of components and communities

3.2.6 Network Evolution Analysis

Like many other social networks, a co-offending network is not a static network and keeps changing over time. Offenders may leave or join the network and their position in the network may change by obtaining or losing power. Links between offenders may form or disappear. Offender groups may appear, split, merge, or disappear. Network structure may change from decentralized to centralized, flat to hierarchical or vice versa. Detecting the evolution patterns of co-offending networks provides important information for law enforcement agencies to understand the behavior of these networks.

We study how a co-offending network evolves over time based on multiple snapshots of the network. For this purpose, we generated five snapshots of the co-offending network for the years 2001–2006. Each snapshot contains the extracted co-offending network from events that happened from 2001 up to that time. For example, $G_a(2004)$ is the co-offending network of all crimes from 2001 to 2004. Below, we examine the evolution of co-offending network based on these five snapshots for various network structural properties.

Figure 3.5 depicts the evolution of size and number of connected components over time. The most interesting observation is that, after 1 year, in the network $G_a(2002)$ there is no large component but it grows in a nearly linear trend. On the other hand, in all networks not many offenders are connected to the medium sized components. The reason is that the medium sized components are merged with the large components through some of their nodes, and we do not have them

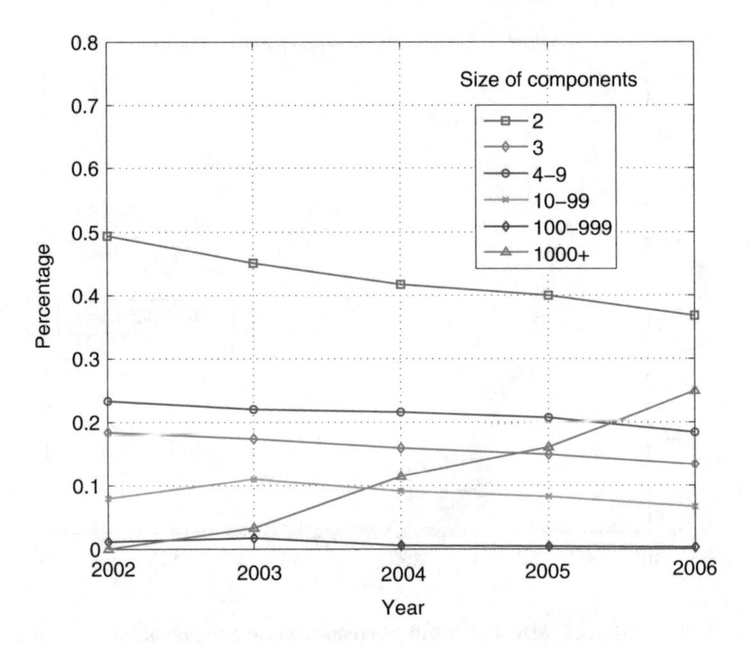

Fig. 3.5 Size and number of connected components vs time

as independent components. In other words, the medium sized components blend in the large components very soon and make them richer; therefore, we do not observe their existence in the network for a long time period. There exists a similar phenomenon in other real-world social networks, a large component tends to merge with the remaining singletons and smaller components [14]. The number of nodes that belong to the small components is almost constant in all 5 years. The reason is that always some of the small components are connected to the medium or large components and simultaneously some new small components appear in the network.

In Fig. 3.6 we plot the evolution of the average distance, diameter, and effective diameter of the co-offending network between 2001 and 2006. This finding may be surprising because of the increasing size of the co-offending network, as network models generally suggest that average distance and diameter should increase with network size [2]. In our case, all these three measures are increasing in the first 3 years and then they start decreasing in the last 2 years. There are studies which report similar results [15].

Figure 3.7 shows how the clustering coefficient changes over the studied time period. There are three observations. First, clustering coefficient is stationary during the 5 years. As expected, clustering coefficient is higher than the expected clustering of a random network with the same number of nodes and edges. Finally, our results are opposite to the empirical studies of some of the social networks [2], where clustering coefficient was found to decrease over time.

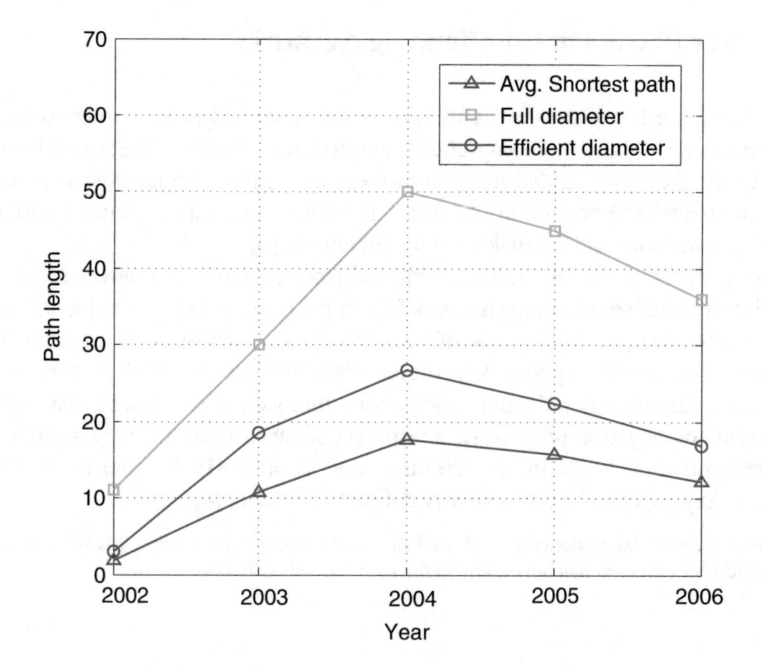

Fig. 3.6 Average distance vs time

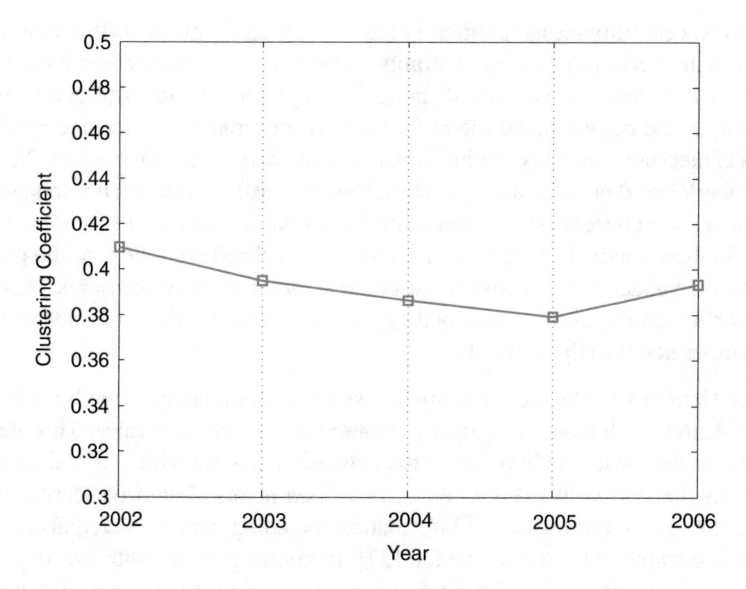

Fig. 3.7 Clustering coefficient vs time

3.3 Key Players in Co-offending Networks

Identifying the key player is a common problem studied in social networks. Key players are potentially more important and also have a higher influence on the other actors [13]. Recognition and removal of these nodes from the network is an aspect of fundamental importance in the study of crime, especially organized crime, for splitting a network and for making it dysfunctional [4].

Key players of a network can be viewed from two different perspectives: their positive or negative role in the network [4]. In the positive key player identification, we need to measure the degrees of connectivity and centrality of an actor in the network, but in the negative key player identification we need to measure the network cohesiveness reduction after nodes removal. Nevertheless, the methods proposed for negative or positive key player identification are very similar with different functionalities. In co-offending network analysis the goal is identifying negative key players. This is formally defined as follows [4]:

> Given a co-offending network, find a set of k nodes such that removing this set of nodes would result in a residual network with the least possible cohesion.

3.3.1 Centrality Measures

Intuitively, centrality measures identify the actors with the greatest structural importance in a network. The existing centrality measures can be divided into three groups based on how they are calculated: node degree, shortest path, and actor ranking methods. Node degree based methods, such as indegree and outdegree measures, are local measures that only use information of the first-level relationships. Methods which work based on information derived from shortest path between actors, such as closeness and betweenness, are considered as global measures. The important point is that in these methods centrality of a node is calculated regardless of the position of the other nodes in the network. In contrast, actor ranking measures, including eigenvector and PageRank, not only they are global, but also they consider centrality of the other nodes in the network.

Degree Centrality Degree centrality is based on the number of outgoing links of an actors. A node with more links obtains greater degree centrality value. This measure focuses on the most visible actors in the network. An actor with a high degree is in direct relationship with many other actors. Such actors should be recognized by other actors as a main channel of information spreading, indeed, a crucial cog in the network, occupying a central position [22]. In contrast, actors with low degree are peripheral in the network and these actors are not active in the connection process. Degree centrality of the actor v is [22]

$$C_D(v) = \frac{|\Gamma_v|}{N-1} \tag{3.7}$$

where $|\Gamma_v|$ is the number of direct neighbors of v, and N is the number of actors in the network.

Closeness Centrality The main idea behind the closeness centrality is that actors that can quickly contact other actors in the network take the central position. The closeness centrality of an actor in a social network is the inverse of the average shortest path distance from the actor to any other actor in the network. This measure shows how much each actor is efficient in spreading information to other actors. The larger the closeness centrality of an actor, the shorter the average distance from the actor to any other actor, and therefore, the better position the actor has in the network. Closeness centrality of the node v is computed as [19]

$$C_c(v) = \frac{N-1}{\sum\limits_{u \in V} d(u,v)} \tag{3.8}$$

where $d(u,v)$ is the distance of node v from node u in the network.

Betweenness Centrality The betweenness centrality is defined as the number of shortest paths between pairs of nodes that pass through the given node. This centrality measure is based on the idea that an actor is key player if it sits in between many other pairs of actors, and it would be traversed by many of the shortest paths connecting pairs of actors. The betweenness centrality of the node v is defined as [8]

$$C_c(v) = \sum_{\substack{u,w \in V \\ u \neq w \neq v}} \frac{\sigma_{uw}(v)}{\sigma_{uw}} \tag{3.9}$$

where $\sigma_{uw}(v)$ represents the total number of shortest paths between each pair of nodes like u and w that pass through node v, and σ_{uw} denotes the total number of shortest paths from u to w.

Eigenvector Centrality The eigenvector method aims to recognize the central actors in terms of the global structure of the network. Eigenvector centrality is defined as the principal eigenvector of the adjacency matrix representing the network. The eigenvector of a network is computed using equation [3]:

$$\lambda v = Av \tag{3.10}$$

where A is adjacency matrix of the network, λ is a constant (eigenvalue), and v is the eigenvector. The idea behind this approach is that actors are central if they have central neighbors. Therefore, centrality of an actor does depend not only on the number of its neighbors, but also on their centrality in the network.

PageRank Centrality PageRank method [6] is a variant of the eigenvector centrality measure which basically is used for ranking the web pages. PageRank models the behavior of a surfer of the web pages, and ranks the web pages based on his behavior. The surfer starts at a random page, and move from a page to another page using the outgoing links. For jumping from a page to another one, the outgoing links are selected uniformly at random. Also the surfer with a probability can jump to any other page. The iteration process is continued until convergence is obtained. This result indicates the chance of a page being visited by the surfer. This method can also be applied on social networks to rank actors. PageRank of the node v is computed as

$$C_P(v) = \frac{1-d}{N} + d\Big(\sum_{\forall u: u \in \Gamma_v} \frac{C_P(u)}{\Gamma_v}\Big) \tag{3.11}$$

where N is the number of nodes in the network, Γ_u is the set of all nodes connecting to u, and d is the probability of continuing the process of moving on the network and not jumping to a random page which is a fixed parameter between zero and one.

3.3.2 Key Players Removal Effects

We believe network centrality analysis can help law enforcement agencies develop strategies for crime reduction and prevention. The current strategy is trivial, from a network perspective: remove those offenders that are most active (nodes with high degree) or commit the most severe crimes. Reiss's argument that some offenders actively recruit new offenders, [18] combined with Liu et al.'s finding that key players (assumed to be the recruiters) are not necessarily the most active criminals in a network [16] warrants a close look at key player identification in co-offending networks. The hiddenness of links and the time-varying structure of these networks necessitate thorough analysis and experimentation to extract the facts to base law enforcement policy on.

In the next section, several experiments were conducted to evaluate the appropriateness of various centrality measures (degree, closeness, betweenness, eigenvector, and PageRank) for identifying important actors in co-offending networks [21]. The crime rate and network structure are intricately linked. The overall crime rate, however, is not equal to the total number of links in the network, since every event involving k offenders translates to a complete graph of size k in our co-offending network. Thus, characteristics of network structure may give some intuition about our original question, a connection which we will revisit in discussing our first experiment.

We investigate the effects of removing central nodes selected using the static network (all two, three, or 4 years worth of data combined) and those selected using a dynamic network (one network for each year). The thought is to account for possible (lack of) persistence in the co-offending network, possibly putting more emphasis on more recent crimes in determining if a person is likely to commit new crimes. Priority given to more recent crimes reflect a real-world bias/predilection/skew of assigning more importance to more recent crimes in determining the criminality of an offender. A small difference between the selections from the static and dynamic networks would lend credibility to the hypothesis that important offenders do not change their game. A larger difference would mean the co-offending network is to some extent transient: offenders cease activity, and other, new offenders start.

All figures below are of offender removal experiments: the top 1 %, 5 %, 10 %, and 20 % of nodes according to each centrality measure is removed from the network, and a line plot of a statistic on the resulting network is shown. A thick dotted line indicates the reference level of the statistic, being either the expected value of the statistic given the number of offender that are removed, or the value of the statistic over the whole network if its change due to node removal cannot be forecast.

3.3.3 *Experiments and Results*

It is common practice that the network in which centrality is measured is the same as the network in which the effects of removal of the most central nodes are measured (see Fig. 3.8, provided here for reference). Most of the results are as one should expect: cutting by degree centrality has the largest effect on average degree, cutting by betweenness has the largest effect on the largest component size, exactly what these centrality measures were designed for. One highlight of these figures is the reasonable efficacy of degree centrality-based node removal, the selection method used by law enforcement agencies, for breaking up the network (Fig. 3.8c), an important feature under our assumption that at least some crime is socially stimulated or facilitated. Other methods (eigenvector and betweenness centrality) can do better, though, partly validating our current research question.

In a crime prevention scenario, and any network that changes over time, this is of course not appropriate: removal happens based on information collected prior to the time of removal, and affects the network after the time of removal. Therefore, we split up the data into five networks, one for each year's worth of data, and tested the effects of intervention (central node removal) after the first year, second year, third year, and fourth year, identifying central nodes in the network of the previous years and removing those from the network of the following years. For reference, we also include the whole network analysis, in which central nodes were identified in the same (whole) network as they were removed from.

In this experiment [21], the top offenders according to each centrality measure computed over the network preceding the intervention were removed from the network at the intervention. Since these offenders are assumed to be dominant actors in the network, causing others to offend, other offenders who only commit

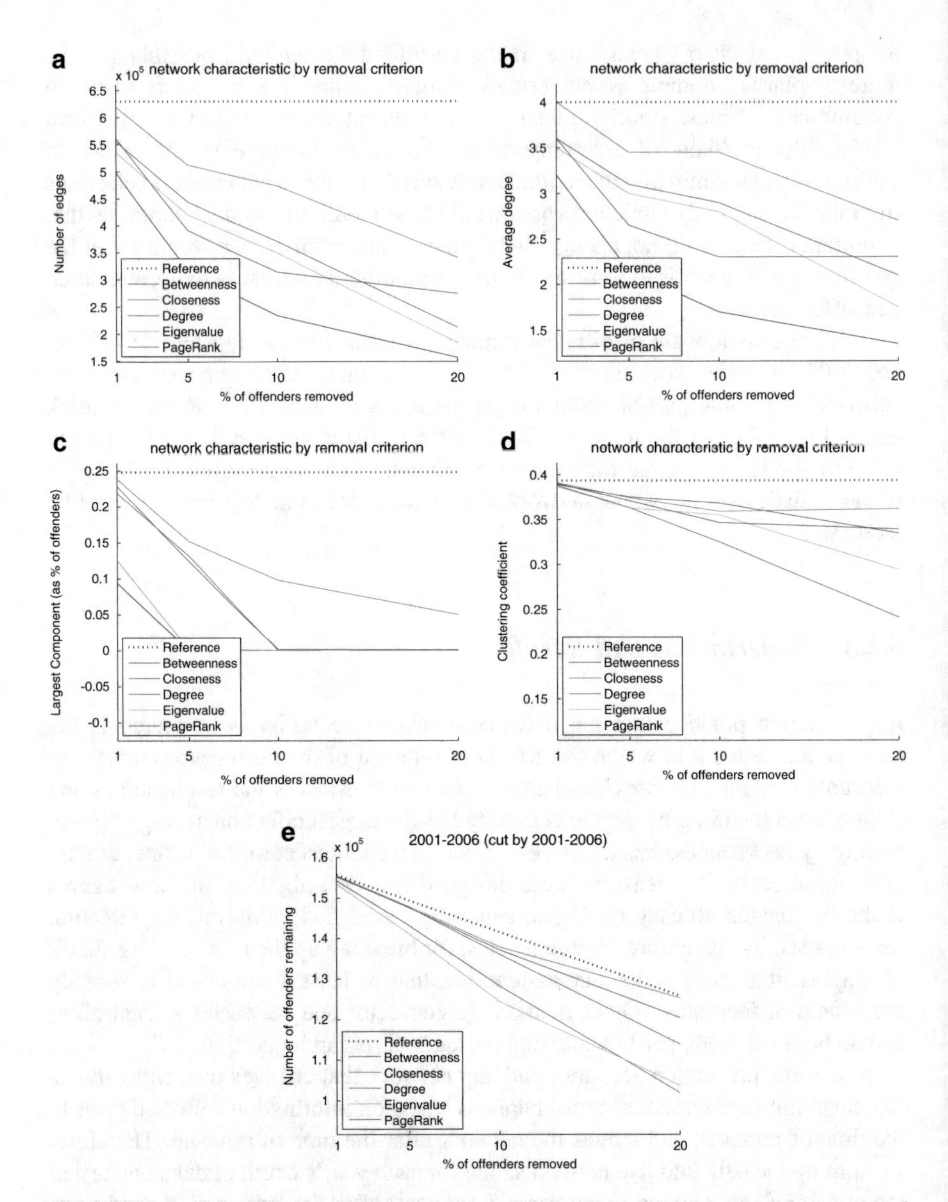

Fig. 3.8 Network statistics for the whole network (2001–2006) after removal of the top *x*% of offenders, according to different centrality measures

crimes with (one of) these top offenders are also removed. The effect on the resulting networks after the intervention is illustrated in Fig. 3.9. Only the number of offenders is reported, but the results for other network statistics are quite similar, showing only marginal change.

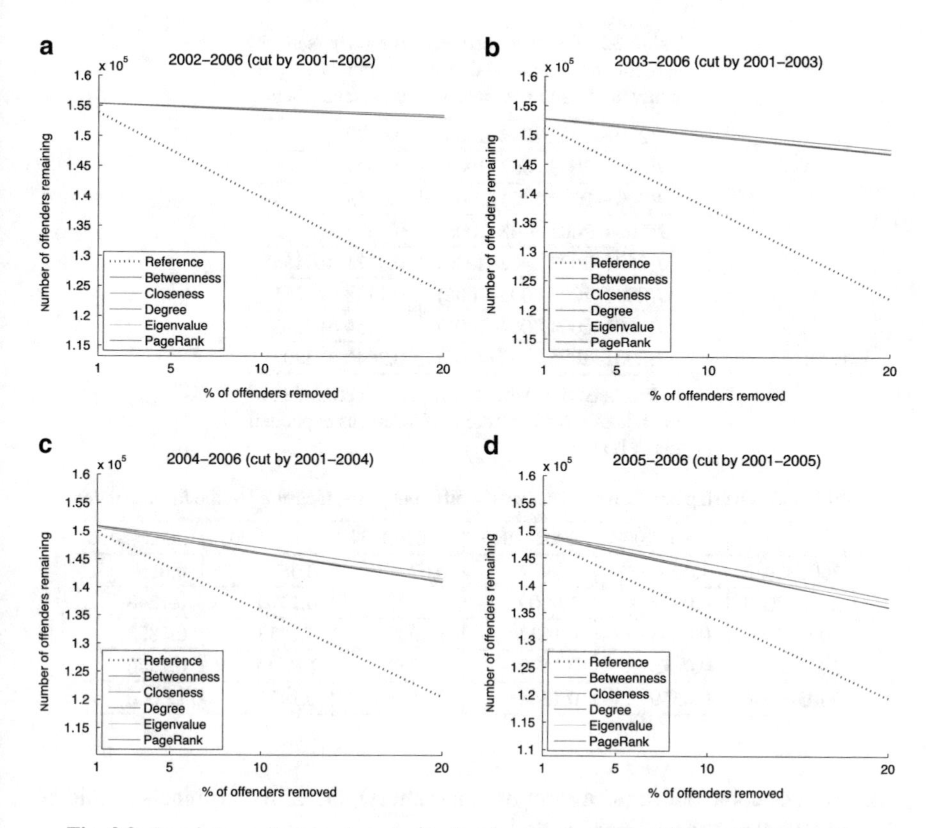

Fig. 3.9 Post-intervention number of offenders for interventions at different points in time

The decrease in the number of offenders in the resulting network is smaller than would be expected (represented by the dashed line). This points to an important feature of the network: transience. Table 3.2 shows the overlap, computed using the Jaccard index

$$J(S_1,S_2) = \frac{S_1 \cap S_2}{S_1 \cup S_2} \tag{3.12}$$

between offender sets of subsequent years, and of pre- and post-intervention networks, and it re-affirms the transient nature of the network.

This transience should be interpreted with caution. One could infer that the majority of offenders come into contact with the police only very infrequently, and this is indeed the case. This does not, however, imply that such offenders are incidental criminals; they may have been arrested, and thereby removed from any opportunity to recur in the data. We currently do not have access to data that would allow us to distinguish between the two, but incidental criminality supposedly being

Table 3.2 Overlap between offender sets in subsequent years, and overlap between pre-intervention and post-intervention offender sets

J(2001–2002,2002–2003)	0.0972
J(2002–2003,2003–2004)	0.1627
J(2003–2004,2004–2005)	0.1853
J(2004–2005,2005–2006)	0.1976
J(2001–2002,2002–2006)	0.0471 (0.0455)
J(2001–2003,2003–2006)	0.1118 (0.1255)
J(2001–2004,2004–2006)	0.1654 (0.2073)
J(2001–2005,2005–2006)	0.2040 (0.2807)

In brackets: the average realized Jaccard index of the sets of most important offenders, as explained in the text

Table 3.3 Overlap between offender sets in different years, measured as the Jaccard index

	2001–2002	2002–2003	2003–2004	2004–2005	2005–2006
2001–2002	1.0000	0.0972	0.0798	0.0675	0.0569
2002–2003	0.0972	1.0000	0.1627	0.1570	0.1248
2003–2004	0.0798	0.1627	1.0000	0.1853	0.1803
2004–2005	0.0675	0.1570	0.1853	1.0000	0.1976
2005–2006	0.0569	0.1248	0.1803	0.1976	1.0000

the result of coincidence (as a form of randomness), incidental criminals should be less likely identified as central nodes.

This supposition is supported by the results presented in Table 3.2, where the average realized Jaccard index is listed in brackets after J(pre-intervention,post-intervention). The realized Jaccard index is the Jaccard index of the selected set of most important offenders of the pre-intervention network and the set of all offenders in the post-intervention network, divided by its theoretical maximum (when all most important offenders recur in the post-intervention network). Table 3.2 shows that, as the amount of historical information included in the selection of important offenders (the pre-intervention network) increases, the selected most important nodes from the pre-intervention network are more likely to occur in the post-intervention network [$p < 0.01$ for J(2001–2005,2005–2006)]. Thus, we have shown that centrality in the co-offending network in the past is at least somewhat informative about criminality in the future, getting back to our original hypothesis.

With the 5 years worth of data batched into individual years, it would seem rather naïve to assign equal importance to events that happened in the first year and events that happened in the last year. This is supported by the overlap between the offender sets of the different years, measured by the Jaccard index in Table 3.3. The overlap between two "time slices" decreases as the amount of time between them increases. This points yet again to the previously mentioned network transience. It also hints at the fact that if we wish to find offenders who will be important after a certain time, we may do well to discount the distant past over the near past.

It may therefore be possible to improve on the previous results by taking into account time, and we compared several time-weighting schemes w:

- none (aggregate network over all years)
- uniform:

$$w_u(C_x, v_i) = \sum_{t=1}^{5} c_x(v_i \in G_t)$$

- linear:

$$w_\ell(C_x, v_i) = \sum_{t=1}^{5} t \cdot c_x(v_i \in G_t)$$

- exponential:

$$w_e(C_x, v_i) = \sum_{t=1}^{5} c_x(v_i \in G_t)^t$$

where $c_x(v_i \in G_t)$ computes centrality x for actor v_i in the network of year t.

If time-weighting changes the set of offenders selected for removal, the structural characteristics of the post-intervention network should also change. The effect on the resulting networks after the intervention is illustrated in Fig. 3.10 as change with respect to the (unweighted) baseline. A positive outcome indicates that the weighting approach was able to further reduce the number of offenders, as compared to the baseline static network. A negative outcome means that the weighting approach actually performed more poorly than without time-weighting. Consistently with the foregoing, only the number of offenders is reported. It is clear that centrality measures measuring more transient features (shortest paths) benefit from taking time into account. Quite counterintuitively, uniform weighting appears to champion the others, suggesting that if historically more important offenders remain important, it does not matter when in history they were important.

3.4 Conclusions

Research in co-offending network mining often lacks access to large real-world crime data sets. One reason for this limitation is the highly sensitive nature of such data and the related privacy issues demanding strict security protocols as well as data storage and processing facilities that meet exceptionally high security standards. An interesting open question is to what extent advanced anonymization techniques can help solving this problem by making secure data more widely available without compromising privacy.

The proposed formal model of crime data and co-offending networks provides a well-defined semantic framework for describing in an unambiguous way the

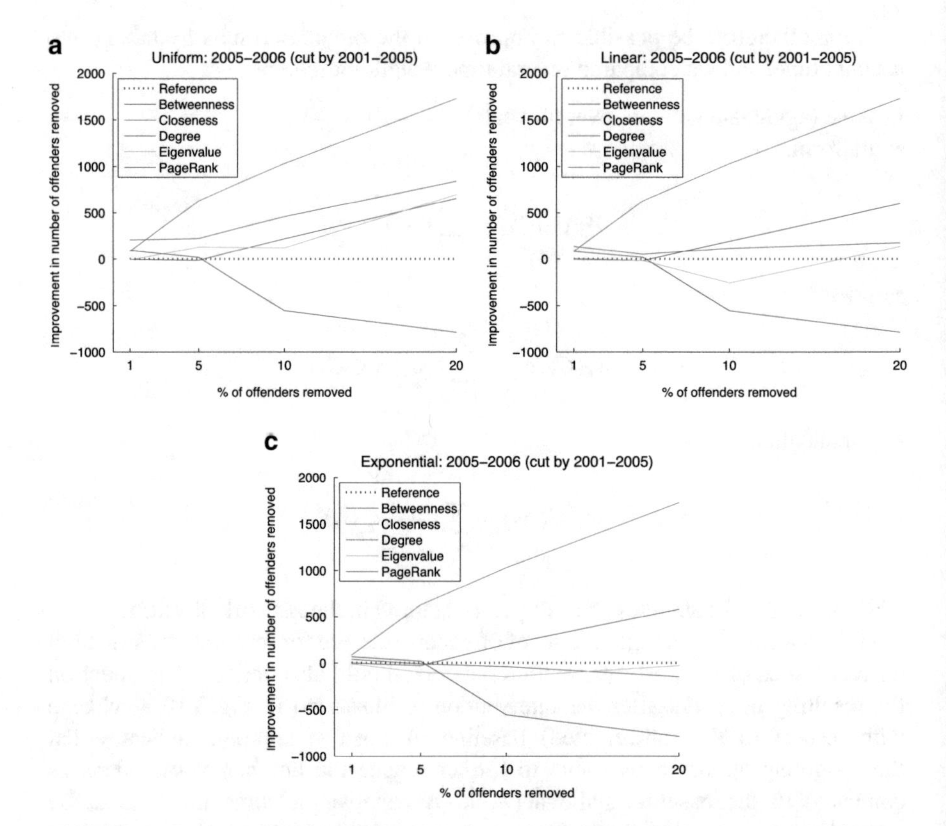

Fig. 3.10 Performance of weighting schemes compared with no weighting scheme (positive values indicate more offenders removed, and hence better performance than unweighted; negative values indicate fewer offenders removed and poorer performance)

meaning of co-offending networks and their constituent entities at an abstract level. Specifically, the formal model aims at bridging the conceptual gap between data level, mining level, and interpretation level, and also facilitates separating the description of the data from the details of data mining and analysis.

In our study we have extracted co-offending networks for a number of most important types of crime, including serious, property, drugs, and moral crimes. The analysis of the co-offending network structure revealed several interesting insights. Surprisingly, the average distance and the diameter of the co-offending network have shrunken in the last few years, indicating a densification of the social network. These results are in line with known studies that show similar phenomena in other types of social networks [15].

Centrality analysis is a well-established field of research in social network analysis, and although its applicability in crime prevention is clear, and its potential impact large, it has not gotten the attention it deserves in recent studies of co-offending networks. This is likely due to the limited accessibility of large-scale data sets, leading researchers to focus on either quite theoretical work [16] or limit their analysis to a relatively small sample [20].

In this chapter we present results of centrality analysis on the BC co-offending network. The ultimate goal is to select offenders such that an intervention entailing their removal would reduce crime rate. Although the efficacy of the investigated centrality measures was limited because of the high transience in the network (offenders ceasing activity and new offenders appearing), we could show that offenders identified as central (by any measure) were more likely to commit further crimes. This effect grew stronger as the amount of information (time observed) used to select central offenders was increased.

Armed with this intuition that time does indeed matter, we compared the performance of centrality measures when computed not over the pre-intervention network as a whole, but over each pre-intervention year individually. We compared several weighting schemes over the centralities per year with unweighted whole-network centrality and found that centrality measures that capitalize on transient characteristics of the network (shortest paths) benefit, whereas centrality measures that rely on more time-stable features of the network did not benefit, or even lost some efficacy. Counterintuitively, equal weighting of all years seemed to be better than discounting the distant past over the near past.

References

1. A.L. Barabasi, R. Albert, Emergence of scaling in random networks. Science **286**(5439), 509–512 (1999)
2. A.L. Barabasi, H. Jeong, Z. Néda, E. Ravasz, A. Schubert, T. Vicsek, Evolution of the social network of scientific collaborations. Physica A **311**, 590–614 (2002)
3. P. Bonacich, Factoring and weighting approaches to status scores and clique identification. J. Math. Sociol. **2**(1), 113–120 (1972)
4. S.P. Borgatti, Identifying sets of key players in a social network. Comput. Math. Organ. Theory **12**(1), 21–34 (2006)
5. P.L. Brantingham, M. Ester, R. Frank, U. Glässer, M.A. Tayebi, Co-offending network mining, in *Counterterrorism and Open Source Intelligence*, ed. by U.K. Wiil (Springer, Vienna, 2011), pp. 73–102
6. S. Brin, L. Page, The anatomy of a large-scale hypertextual web search engine. Comput. Netw. ISDN Syst. **30**(1), 107–117 (1998)
7. G. Bruinsma, W. Bernasco, Criminal groups and transnational illegal markets. Crime Law Soc. Chang. **41**(1), 79–94 (2004)
8. P.J. Carrington, J. Scott, S. Wasserman, *Models and Methods in Social Network Analysis* (Cambridge University Press, New York, 2005)
9. A. Clauset, C.R. Shalizi, M.E.J. Newman, Power-law distributions in empirical data. SIAM Rev. **51**(4), 661–703 (2009)
10. E.W. Dijkstra, A note on two problems in connexion with graphs. Numer. Math. **1**(1), 269–271 (1959)
11. P. Erdös, A. Rényi, On random graphs, I. Publ. Math. Debr. **6**, 290–297 (1959)
12. M. Girvan, M.E.J. Newman, Community structure in social and biological networks. Proc. Natl. Acad. Sci. **99**(12), 7821–7826 (2002)

13. D.M.A. Hussain, D. Ortiz-Arroyo, Locating key actors in social networks using bayes posterior probability framework. Intell. Secur. Inform. **5376**, 27–38 (2008)
14. R. Kumar, J. Novak, A. Tomkins, Structure and evolution of online social networks, in *Proceedings of the 12th ACM SIGKDD International Conference on Knowledge Discovery and Data Mining (KDD'06)* (2006), pp. 611–617
15. J. Leskovec, J. Kleinberg, C. Faloutsos, Graph evolution: densification and shrinking diameters. ACM Trans. Knowl. Discov. Data **1**(1), 2 (2007)
16. X. Liu, J. Bollen, M.L. Nelson, H. Van de Sompel, Co-authorship networks in the digital library research community. Inf. Process. Manag. **41**(6), 1462–1480 (2005)
17. C.R. Palmer, P.B. Gibbons, C. Faloutsos, Anf: a fast and scalable tool for data mining in massive graphs, in *Proceedings of the 8th ACM SIGKDD International Conference on Knowledge discovery and Data Mining (KDD'02)* (2002), pp. 81–90
18. A.J. Reiss Jr., Co-offending and criminal careers. Crime Justice **10**, 117–170 (1988)
19. G. Sabidussi, The centrality index of a graph. Psychometrika **31**(4), 581–603 (1966)
20. M.K. Sparrow, The application of network analysis to criminal intelligence: an assessment of the prospects. Soc. Netw. **13**(3), 251–274 (1991)
21. M.A. Tayebi, L. Bakker, U. Glässer, V. Dabbaghian, Locating central actors in co-offending networks, in *Proceedings of the 2011 International Conference on Advances in Social Networks Analysis and Mining (ASONAM'11)* (2011), pp. 171–179
22. S. Wasserman, K. Faust, *Social Network Analysis: Methods and Applications* (Cambridge University Press, New York, 1994)

Chapter 4
Organized Crime Group Detection

In this chapter, we propose a new computational approach to organized crime group detection based on a social network analysis perspective. A challenging aspect is the need for a precise definition of *what* exactly constitutes a criminal organization. Confronted with a bewildering diversity of characteristics in definitions of organized crime and criminal organizations, the conceptual model of organized crime appears not clearly rendered in the literature—at least not for the purpose of computational analysis [3, 6, 8, 10, 18, 20].

Striving for a general and open definition, a natural source is the criminal code, although this depends on the specific country. Starting from the definition of criminal organization in the *Criminal Code of Canada* [10], we focus on methodical and analytical aspects in utilizing social network analysis methods for organized crime group detection. The main goal is to promote co-offending network analysis as an effective means for extracting information about criminal organizations from police-reported crime data. We contend that it would be virtually impossible to obtain such information by using traditional crime analysis methods.

The approach described here comprises three major building blocks: (1) a group detection method, an extension of the clique percolation method [14], to match a working definition of offender group; (2) a crime assessment method which covers and formalizes common characteristics of organized crime found in the criminology literature; (3) a group evolution model for analyzing offender group behaviour over the observable life cycle of a group.

Section 4.1 discusses related work. Section 4.2 introduces the concepts and the terminology, and the problem definition. Section 4.3 presents the organized crime group detection framework. Next, Sect. 4.4 discusses our experimental results on the BC crime dataset. Section 4.5 concludes this chapter.

M.A. Tayebi, U. Glässer, *Social Network Analysis in Predictive Policing*,
Lecture Notes in Social Networks, DOI 10.1007/978-3-319-41492-8_4

4.1 Background

Studying prominent historical research on how organized crime developed in New York City, Block [3] concludes that "organized crime was not only more fragmented and chaotic than believed, but also it involved *webs of influence* that linked criminals with those in positions of power in the political and economic world." He argues that these patterns of affiliation and influence were far more important than any formal structure, since they allowed criminals to maximize opportunities, and should be considered a social system.

The social system of organized crime [3]:

> *...refers to the notion that organized crime is a phenomenon recognizable by reciprocal services performed by professional criminals, politicians, and clients. Organized crime is thus understood to lie in the relationships binding members of the underworld to upperworld institutions and individuals. Organized crime is not a modern, urban, or lower-class phenomenon; it is a historical one whose changes mirror changes in civil society, the political economy. That is why, naturally, organized crime is increasingly taken to represent a series of relationships among professional criminals, upperworld clients and politicians ...*

Confronted with a bewildering diversity of characteristics referred to in existing definitions of organized crime and criminal organizations, the conceptual model itself appears not clearly rendered in the literature. Striving for a definition that is general and open, a potential source is the criminal code, although this depends on a specific country. For instance, a baseline definition of criminal organization is provided by the Criminal Code of Canada [10, p. 49]:

> *In Canada a criminal organization is a group, however organized that: (a) is composed of three or more persons in or outside Canada; and (b) has as one of its main purposes or main activities the facilitation or commission of one or more serious offences, that, if committed, would likely result in the direct or indirect receipt of a material benefit, including a financial benefit, by the group or by any one of the persons who constitute the group. The definition further specifies that it excludes a group of three or more persons that has formed randomly for the immediate commission of a single offence. Section 467.1(1) of the Criminal Code of Canada.*

Looking for a quantitative definition, in an attempt to measure organized crime, van der Heijden proposes a number of common characteristics [18]:

1. *Collaboration of more than two people;*
2. *Commission of serious criminal offences (suspected);*
3. *Determined by the pursuit of profit and/or power;*
4. *Each having their own appointed tasks;*
5. *For a prolonged or indefinite period of time;*
6. *Using some form of discipline and control;*
7. *Operating across borders;*
8. *Using violence or other means suitable for intimidation;*
9. *Using commercial or businesslike structures;*
10. *Engaged in money laundering;*
11. *Exerting influence on politics, the media, public administration, judicial authorities, or economy.*

According to [18], for any criminal group to be categorized as organized crime it needs to have at least six of the above characteristics, where items 1, 2, and 3 are obligatory, thus adding three more characteristics.

A major study in the Netherlands [8] mentions great variations in collaborative forms of organized crime and concludes that "the frameworks need not necessarily exhibit the hierarchical structure or meticulous division of labor often attributed to mafia syndicates. Intersections of social networks with a rudimentary division of labor have also been included as groups in the sub-report on the role of Dutch criminal groups, where they are referred to as cliques. As is demonstrated ... there can be sizable differences in the cooperation patterns within these cliques and between the cliques and larger networks of people they work with on an incidental basis."

An impressive collection of definitions of organized crime specific for various countries, comprising more than individual 150 entries in total, has been gathered by von Lampe [20]. In addition, this collection also includes comments on how to define organized crime, and definitions by prominent individuals and government agencies, for instance, such as the Federal Bureau of Investigation (FBI). Not included though are definitions of the term "organized crime group." Given the abstract nature and informal language of these definitions, it is not clear at all how and to what extent one may utilize this resource for defining organized crime in precise computational and/or mathematical terms.

In most cases, existing definitions in the literature on organized crime concentrate on three essential perspectives for characterizing the nature of this form of crime [19]: In the first view, organized crime is primarily about crime. Organized crime is seen as a *specific type* of criminal activity that has certain specific characteristics such as continuity in contrast to irregular criminal behaviour. In the second view, organized crime is more related to the *concentration of power*, either in economic or in political structures of the society. And in the third view, the emphasis is on *organized*. That is, the important aspect of organized crime is on how offenders are connected to each other more than what they do.

Based on the third view, we formalize the central aspects of criminal networks in a coherent and consistent formal framework to provide a precise semantic foundation that is consistent with criminological research, social network analysis, and law enforcement operations. Our work aims at bridging the conceptual gap between data level, mining level, and interpretation level, and is intended for developing advanced computational methods for analyzing co-offending networks to detect and extract organized crime structures and how they evolve over time in order to assist law enforcement and intelligence agencies in their investigations.

4.1.1 Community Detection in Social Networks

This section addresses the concept of community in social networks and explores community detection methods. Crime groups in co-offending networks can be viewed as communities in social networks and thus may be identified using existing detection methods.

Community detection in social networks has attracted considerable interest and many definitions of the concept of community have been proposed. In social science studies, social networks are considered as basis of social behaviours and activities. Studies of different social networks show that community structure influences information transfer, communication, and cooperation. Sense of community is generally defined as a feeling that members of a group matter to one another and to the group, and a common belief that members' needs will be satisfied through their commitment to be together [12].

Community Detection in Static Networks Algorithms for community detection in static graphs are usually looking for a "good" partition of the nodes. The main problem is "what does 'good' really mean?". Community detection methods can be divided into three types:

Node-Based Methods. In these methods each node in a group needs to have specific properties, meaning that two nodes are considered as members of same community if their positions in the network satisfy some specific constraints. Clique is an example for these types of communities, which works based on complete mutuality in the group. Clique is a maximal complete subgraph of nodes in which every two nodes are connected by an edge.

It is NP-complete to identify the maximum clique in a graph. One brute-force approach to find the maximal clique is to scan all nodes in the network. Then, for each node the maximal clique including that node is detected. Each node is removed from the network when it is traversed once. This method works for small-scale networks, but it is impractical for large-scale networks. The main strategy to address this issue is to efficiently prune those nodes that are unlikely to be included in a maximal clique.

In the proposed approach by Abello et al. [1] in each step a subset of network is sampled. In this subset the maximal clique is identified using a greedy search approach. The detected clique is used as the lower bound of pruning in the next steps. If the largest maximal clique up to this step has p nodes, nodes with degree less than p will be removed in the next steps.

There are other node-based methods which consider the reachability between nodes. For instance, k-clique is a subgraph that the length of shortest path among any two nodes is not greater than k, or a k-clan is defined as a k-clique that the length of shortest path between any two nodes in the subgraph is not greater than k. The difference of k-clique and k-clan is that the shortest path length in the k-clique is defined on the original network, but for the k-clan it is on the subgraph.

There are also node-based approaches that need each community member be neighbor of specified number of other members. For example, k-plex is a subgraph with n nodes, that each node has at least $n-k$ neighbors. Generally such methods are not suitable for community extraction from real-world social networks where the nodes degree distribution is usually power-law.

The other constraint for a community can be defined as having more connections to the other nodes inside the community than to those outside the community. For example, LS set [4] is defined as a set of nodes C that any of its proper subsets has more links to its complement within C than outside C.

Group-based methods. In the second class of methods for community detection the focus is on the structure of the connections inside the groups. On contrary to the node-based methods, in group-based methods nodes in the group can have low connectivity as long as all nodes satisfy some defined criteria. One example of such criteria is density of the edges. In the density based groups if the density of the edges in a group of nodes is bigger than a threshold then that group of nodes is called a community.

Network-based methods. In the network-based methods the goal is partitioning the network to some disjoint groups by optimizing some criteria. For dealing with this problem, some quality measures have been defined that give a score to a partition. In the network-based methods a good partition is one which maximizes this quality measure. One of the most commonly used quality measures is *modularity* [13], and maximizing modularity in a greedy manner is one of the predominant methods for community detection. Modularity Q is defined as

$$Q = \sum_i (e_{ii} - a_i^2) \tag{4.1}$$

where e_{ij} is the fraction of edges that connect nodes in community i to nodes in community j, and $a_i = \sum_j e_{ij}$. But it has been shown that modularity maximization is an NP complete problem [5], and thus most of the solutions for this problem are based on approximation algorithms.

Community Evolution Tracking In studies of how communities evolve over time, two main approaches have been used: (1) applying temporal information directly in the community detection process, and (2) tracking communities over a number of snapshots in time.

To take into account temporal information, recently, a new type of clustering, called *evolutionary clustering*, that captures the evolutionary process of clusters in time-stamped data was introduced. Chakrabarti et al. [7] address this issue in their paper by proposing a framework called *temporal smoothness*. The output of this framework is a sequence of clusterings, one for each timestep by considering two distinct aspects: first, it should have low *history quality*, which means it should be similar to the previous clustering in the sequence, and, second, it should have high *snapshot quality* which means it should have high accuracy in clustering of the current arrived data. The evolutionary clustering algorithm takes the similarity matrices $M_1, M_2, \ldots, M_t$ and produces the clusterings $C_1, C_2, \ldots, C_t$. Evolutionary clustering uses a cost function to trade off the history quality and the snapshot quality. The cost function consists of two parts comprising snapshot cost and temporal cost

$$\alpha \times \mathrm{SC}(C_t, M_t) + (1-\alpha) \times \mathrm{TC}(C_{t-1}, C_t) \tag{4.2}$$

where the parameter α $(0 < \alpha < 1)$ is used to adjust the level of preference for each of the two costs accordingly.

In the cost function snapshot cost measure SC, the quality of clustering C_t at time t, with respect to M_t and temporal cost TC, determines how similar the current clustering C_t is compared with the previous clustering C_{t-1}. For the snapshot and temporal cost, the relation is the smaller the value, the better the quality. The temporal smoothness framework attempts to find a clustering C_t that minimizes Eq. (4.2).

Several evolutionary graph clustering methods have been proposed under the temporal smoothness framework, for instance FacetNet [11], which extended the soft clustering algorithm [21] from static graphs to dynamic graphs.

Another method for identifying relations between communities is constructing the networks for each time step. First, communities are identified within each of these networks, then relationships among communities on subsequent snapshots are recognized. Hence, such an algorithm operates in two steps: (1) static community detection on each snapshot, and (2) applying a matching function to recognize how these static communities evolve over a number of time steps. We use this approach for identifying organized crime groups evolution trace, as illustrated in Sect. 4.3.2.

4.2 Concepts and Definitions

This section introduces the basic concepts and definitions used in the following sections [17]. We define the concept of *offender group* as a basic substructure of a co-offending network and describe the analytic method for tracing how offender groups evolve over their "lifecycle." Finally, we define the concept of *organized crime group* in terms of characteristics that discriminate possible criminal organizations from regular offender groups. The rationale for the applied characteristics is to be in line with the respective definition in the *Criminal Code of Canada*.

Offender Group An offender group comprises three or more offenders who collaborate in committing offences. This does not mean that each and every group member participates (in an active role) in all offences committed. These groups are not necessarily formed as the result of a predefined plan and also they need not be active continuously. Their members have generally local clustering within larger loosely connected networks, thus constituting a small group with varying degrees of connection to other larger groups. In our model, $C_1^t, C_2^t, \ldots, C_n^t$ refer to n offender groups in the co-offending network at time period t.

Group Activity For offender group C_i^t, the *activity* $\Theta_i^{t_1,t_2}$ states how frequently members of this group have committed offences during time period t_1 compared to time period t_2.

Group Criminality Group criminality Φ_i^t represents a measure for the degree of *seriousness* of offences committed by members of offender group C_i^t during time period t.

Active Offender Group Active offender groups have a history of continued criminal activity over some longer time period. $A_i^{t_1,t_2}$ represents an active offender group that is active at time period t_1 and is still active at time period t_2.

Serious Offender Group An offender groups whose overall criminal activity at time period t shows a high degree of serious criminal offences is called serious offender group and is referred to by S_i^t.

In theory, the two concepts of organized crime group and offender group differ in at least three basic aspects: (1) Group scale and motivation; (2) Time interval of collaboration; and (3) Type of criminal activity. In practice, however, the distinction between organized crime group and offender group is not always clear-cut and can be challenging. To qualify as criminal organization, a necessary (but not sufficient) condition is the commission of serious offence motivated by material benefit. While the meaning of "serious offence" can be clearly defined in terms of offences classified as indictable or hybrid offence or statute serious offence in the *Criminal Code / Controlled Drugs and Substances Act*, the meaning of material benefit may be interpreted in a narrow or in a broader sense.

4.2.1 Problem Definition

Problem 1: Organized Crime Group Detection Given $G_t, G_{t_1}, \ldots, G_{t_n}$, the co-offending networks extracted for the time intervals $t, t_1, \ldots, t_n$, and α and β as group activity and criminality thresholds, the goal is detecting organized crime groups $O_1^{t_i}, O_2^{t_i}, \ldots, O_m^{t_i}$ embedded in the co-offending network at time period t_i.

Problem 2: Organized Crime Group Evolution Trace Given detected organized crime groups in the time intervals $t, t_1, \ldots, t_n$, the goal is identifying the evolution trace of O_a^t, $E(O_a^t)$, which is a sequence $O_a^t, O_{a_1}^{t_1}, O_{a_2}^{t_2}, \ldots, O_{a_n}^{t_n}$ of related organized crime groups over n consecutive time periods that shows the dynamic transformation, or evolution, of the organized crime group O_a^t since time period t.

4.3 Proposed Approach

In this section we first present the proposed approach for organized crime group detection, and second, we describe the model for tracing organized crime group evolution [17].

4.3.1 Organized Crime Group Detection

Community detection is a prominent research topic in social networks. The nature of organized crime groups, however, is different from other types of groups like friendship or co-authorship groups. Organized crime groups are usually well established with group membership being defined explicitly and strictly. Unlike friendship or co-authorship communities, offender groups as well are characterized by member relationships that are more systematic and organized to achieve material benefit from committing crime. Therefore, detecting organized crime groups calls for a stricter definition of community.

Based on fundamental discussions in the criminology literature, one can summarize the important characteristics of organized crime groups as: (1) These groups have at least three members and can be categorized as centralized or distributed or hierarchical groups. Regardless of this classification, our focus is on offender groups for which the density of their intra-group collaborations is higher than the density of intergroup collaborations; (2) Organized crime groups are characterized by a distribution of roles and different degrees of agency amongst individuals, where groups can overlap and may have common members; (3) These groups commit serious crimes with the perspective of gaining material benefit; (4) Their activity is more continuous compared to regular offender groups.

For the purpose of organized crime group detection, in each time snapshot of a co-offending network the following tasks are carried out in consecutive steps: (1) Discover offender groups in the current network; (2) Compute the activity and criminality of these groups in the time period between the current network and the previous network based on the offences that were committed by their members; (3) Assess the material benefit associated with each of the offences considered in Step 2; (4) Identify those groups that qualify as possible criminal organizations; (5) Update the groups evolutionary trace for the current time period. In the following each of these steps is explained in more detail.

Offender Group Detection In the first step of the proposed method, offender groups are built up from k-cliques. A group consists of adjacent k-cliques, sharing at least $k-1$ nodes with each other. Since an offender group should have at least three members, we assume $k = 3$. Each clique uniquely belongs to one community, but cliques within different communities may share nodes. Hence, we have overlapping groups with common members. For each offender group C_i, these members are assigned as their kernels $K(C_i)$. Kernels are the main members of an offender group and are completely involved in the group activities. In the second step, neighbor nodes connected directly to the kernels are added to the offender groups. These nodes are called *peripheries*. Peripheries of an offender group C_i are denoted by $P(C_i)$.

Organized Grime Group Detection Activity and criminality of an offender group are two key characteristics toward understanding the group structure. Below we present how these two measures are computed. Criminality of an offender group C_i at time step t, denoted by $\Phi_t(C_i)$, is defined as

$$\Phi(C_i) = \sum_{k=1}^{k=n} \frac{\varphi_{i_k}}{n} \tag{4.3}$$

where φ_{i_k} indicates the seriousness of an offence i_k that is committed by members of group C_i at time step t. Let $i_1, i_2 \ldots, i_n$ be the offences in which members of C_i were involved at time step t. Activity of the offender group C_i at time t_1 with respect to time t_2, denoted by $\Theta_{t_1,t_2}(C_i)$, is computed as follows:

$$\Theta_{t_1,t_2}(C_i) = \frac{|R_{t_1}(C_i)|}{|R_{t_2}(C_i)|} \tag{4.4}$$

where $|R_{t_1}(C_i)|$ and $|R_{t_2}(C_i)|$ denote the number of binary relationships (co-offences) within offender group C_i at time steps t_1 and t_2, respectively.

In order to determine whether a detected offender group qualifies as a possible organized crime group, activity and criminality of the group are considered. For this purpose, we define two thresholds α-*activity* and β-*criminality*. A given offender group C_i is considered as active group A_i, if $\Theta(C_i) > \alpha$, and as serious group, if $\Phi(C_i) > \beta$. We consider an offender group a possible organized crime group, if it is serious and active. Meaningful values for α and β are to be determined experimentally. Algorithm 1 outlines the pseudo-code of this approach.

Algorithm 1 Organized crime group detection

Require: (1) Crime event dataset
(2) Crime seriousness index
(3) Activity and criminality thresholds α, β
Ensure: Organized crime groups $O^t_1, O^t_2, \ldots, O^t_m$

1: /* Data Preparation */
2: **For each** set of crime incidents in the interval $[t_1, t_2]$
3: Extract the co-offending network
4: Detect offender groups $C^t_1, C^t_2, \ldots, C^t_n$
5: **For each** offender group $C^t_i \in C^t$
6: Compute the group *activity* Θ^{t_1,t_2}_i
7: Compute the group *criminality* Φ^t_i
8: Identify possible organized crime groups
9: **For each** possible organized crime O^t_i
10: Assess overall group *material benefit*
11: Apply the evolution trace model on $O^t_1, O^t_2, \ldots, O^t_m$

4.3.2 Organized Crime Group Evolution Model

Like other communities, organized crime groups typically evolve over time. An organized crime group may grow by admitting new members, shrink by losing members, split into smaller groups, or a new group may form by merging existing groups. Therefore, we devise a model that addresses all these aspects of organized crime group evolution.

The model needs to determine which group at previous time has evolved into which group at current time. Five phenomena can occur for a group in a single snapshot: a community may survive, split, merge, emerge, or cease [16]. For this purpose we introduce a matching function

$$\text{match} : \mathscr{G} \times 2^{\mathscr{G}} \rightarrow \mathscr{G}$$

where $\mathscr{G}$ denotes a set of groups and $2^{\mathscr{G}}$ is the powerset of $\mathscr{G}$. For a given organized crime group O_i^t and set of organized crime groups $\mathscr{G}^{t_1}$, let $\text{match}(O_i^t, \mathscr{G})$ yield the group $O_i^{t_1}$ such that this group has the largest intersection with O_i^t above a given threshold λ, as formally defined below.

$$\begin{aligned} \text{match}(O_i^t, \mathscr{G}) = O_j^{t_1} \ \text{ with } \forall\, O_k^{t_1} : O_k^{t_1} \in \mathscr{G} \wedge \\ \text{overlap}(O_i^t, O_j^{t_1}) \geq \text{overlap}(O_i^t, O_k^{t_1}) \wedge \\ \text{overlap}(O_i^t, O_j^{t_1}) > \lambda \end{aligned} \tag{4.5}$$

where, for two organized crime groups $O, O' \in \mathscr{G}$, we define

$$\text{overlap}(O, O') = \min\left(\frac{|O \cap O'|}{|O|}, \frac{|O \cap O'|}{|O'|}\right) \tag{4.6}$$

Using the matching function, we apply the following rules for tracking the evolution of organized crime groups:

- O_i^t survives in the next time slot as $O_j^{t_1}$, if $O_j^{t_1} = \text{match}(O_i^t, \mathscr{G}^{t_1})$ and for each $O_k^t \neq O_i^t$, $O_j^{t_1} \neq \text{match}(O_k^t, \mathscr{G}^{t_1})$.
- O_i^t splits into groups $O_1^{t_1}, O_2^{t_1}, \ldots, O_n^{t_1}$, if there is enough overlap between each of these splitted groups and O_i^t, and also $(O_1^{t_1} \cup O_2^{t_1} \cup \ldots \cup O_n^{t_1}) \cap O_i^t$ is above a predefined minimum defined threshold.
- O_i^t merges with some other groups into $O_j^{t_1}$, if $O_j^{t_1} = \text{match}(O_i^t, \mathscr{G}^{t_1})$ and $\exists\, O_k^t \neq O_i^t$: $O_j^{t_1} = \text{match}(O_k^t, \mathscr{G}^{t_1})$.
- O_i^t ceases, if none of the above scenarios happened.
- $O_j^{t_1}$ emerges, if $\forall\, O_i^t$: $O_j^{t_1} \neq \text{match}(O_i^t, \mathscr{G}^{t_1})$.

These rules are intuitive and easy to observe in the life cycle of groups, but they are not yet rigorous enough. The main problem lies in defining the threshold λ. This threshold needs to be determined based on experimentation and observation, for instance, by learning from existing histories for real-world organized crime groups.

4.4 Experiments and Results

This section describes the experimental evaluation of the proposed method for organized crime group detection on the BC crime dataset [9]. A basic principle in the characterization of "criminal organization" defined in the Criminal Code is "the facilitation or commission of one or more serious offences, that, if committed, would likely result in the direct or indirect receipt of a material benefit ... by the group or by any one of the persons who constitute the group." Consequently, the specific type of committed offences plays a crucial role in determining whether or not an identified crime group is considered a possible organized crime group. In abstract computational terms this aspect effectively constrains the search space to be analyzed by the applied data mining algorithms, such that all offences that do not qualify as a serious offence with associated material benefit are disregarded. In general, knowledge discovery in databases and data mining is ultimately restricted to the information encoded in the underlying datasets. This experiment was run on historic anonymized data with limitations in quality (referring to attributes such as completeness, consistency, and noise) as in all historic police datasets where verification against the "grand truth" is often not feasible or even impossible so one has to work with the data as it is. In practice, data mining algorithms, when used with current and enhanced crime datasets, provide decision support for exploring hand-entered data.

4.4.1 Offender Groups Characteristics

In this part we explore the characteristics of the extracted offender groups and active offender groups. The crime data is partitioned into the following five time snapshots, each of which represents a 12-month time interval: Mid-2001–Mid-2002; Mid-2002–Mid-2003; Mid-2003–Mid-2004; Mid-2004–Mid-2005; and Mid-2005–Mid-2006. Figure 4.1 shows the number of offender groups for different clique sizes k. As expected, the number of offender groups decreases with increasing the clique size. All experiments discussed below are based on clique size k equal or greater than 3, the minimum group size required for a criminal organization under the Criminal Code.

As a simple example, we assume an offender group C_1^2 consisting of seven members with ten co-offending links detected at time snapshot $t = 2$. We follow the behaviour of this group at time snapshots $t = 3$, $t = 4$ and $t = 5$, and further assume that in the snapshots 3–5 we have observed 2, 5, and 9 co-offences, respectively, among members of group C_1^2. Then, the computed activity for these snapshots would

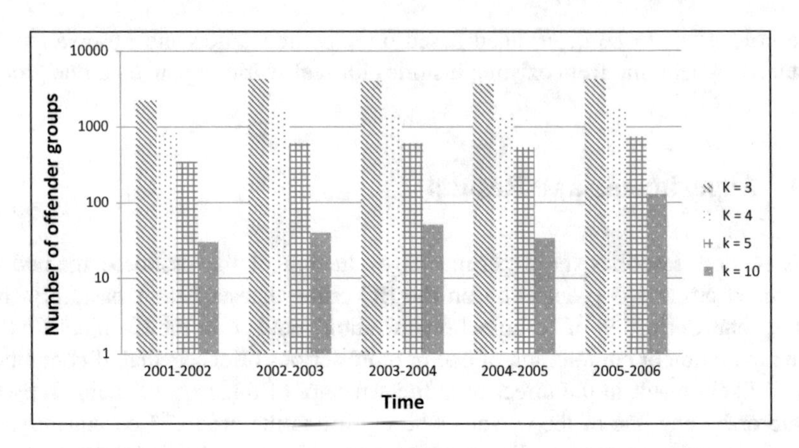

Fig. 4.1 Number of offender groups using different minimum clique size

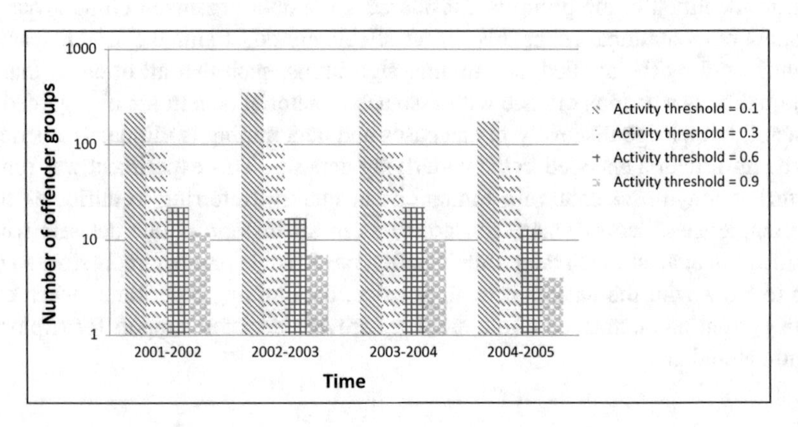

Fig. 4.2 Number of observed offender groups by different activity thresholds

be $\Theta_1^{2,3} = 0.2$, $\Theta_1^{2,4} = 0.5$ and $\Theta_1^{2,5} = 0.9$. Activity measures the relative activity of a co-offending group as observed in a particular time snapshot, compared to the number of distinct co-offences of that same co-offending group in the snapshot this group was first detected. Threshold values are introduced as a flexible means to adjust the level of observed activity above which co-offending behaviour is actually taken into account for the analysis. Implicitly, higher levels of activity observed over a number of consecutive time snapshots suggest a higher degree of stability of a co-offending group. Figure 4.2 illustrates the number of offender groups in each time snapshot for different *activity thresholds* α. An *activity threshold* α describes the percentage of the structure of the co-offending group that remains unchanged between time snapshots.

Even with 60 % of the structure of the co-offending group remaining intact between snapshots ($\alpha = 0.6$) still about 1 % of all the offender groups remain in the list of active offender groups, meaning that some offender groups keep their collaboration intact and unchanged over longer time periods.

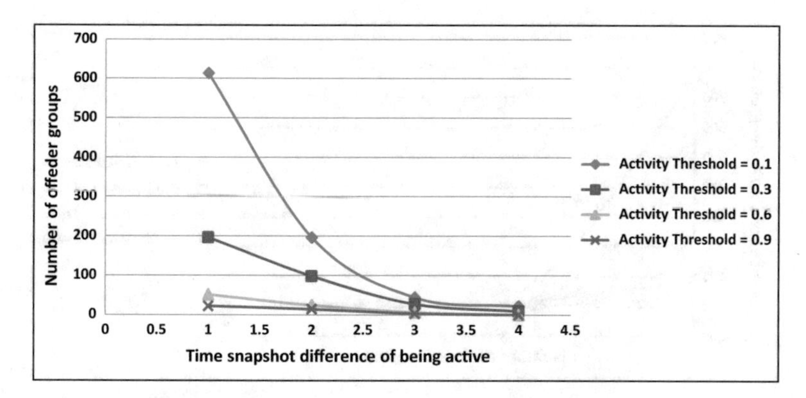

Fig. 4.3 Number of offender groups active over a range of time snapshots (which may include inactive snapshots other than the first and the last snapshot in that range)

Active offender groups can be further characterized as continuously active over several consecutive snapshots or as sporadic, with their activity occurring at irregular time intervals with inactive intermediate snapshots.[1] In assessing the continuity of activities of offender groups, we study their criminal activity over several time snapshots, applying activity thresholds. For instance, assume the following scenario. Group C_1^3 was detected at snapshot $t = 3$, and this group has four members and four co-offending links. No activities were observed in time snapshot $t = 4$. At snapshot $t = 5$, three co-offences were observed. For any activity threshold equal to or less than 0.75, we can consider the time snapshot difference for group C_1^3 equal to two.

Figure 4.3 shows the number of offender groups observed over time periods with one, two, three, and four years difference. The important point here is that with increasing time difference the number of observed groups decreases exponentially. Even with very low values for α only few groups can be observed over four snapshots, and with high values for α no group can be observed over four snapshots. However, one can also see that from one snapshot to the next one continued group activity is more common, even for higher values of α. This finding supports the theory of short-time collaborations of most offender groups.

According to [2], many criminal organizations are short-lived and comprised of offenders with desired skills who form temporary networks to take advantage of a crime opportunity. It is concluded [2] that these groups often dissolve after exploiting the opportunity, looking for new chances which may need other combinations of skills.

---

[1] We are aware of the possibility that apparent sporadic activity could result from group activities not coming to the attention of the police during a given time period rather than from actual lack of criminal activity during that time period.

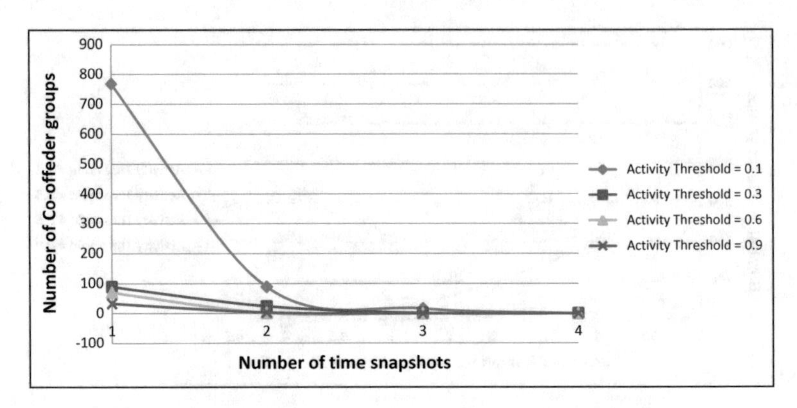

Fig. 4.4 Number of offender groups by the total number of observed time snapshots in which a group was considered active

Another important aspect of active offender groups is the number of snapshots in which a group is being active. Assume an offender group C_1^1 having 4 members and 5 co-offending links detected at time snapshot $t = 1$. We have observed 1, 0, 4, and 5 co-offences, respectively, among members of this group in the time snapshots from $t = 2$ to $t = 5$. Activity of this group for each of these snapshots would be, respectively, $\Theta_1^{1,2} = 0.2$, $\Theta_1^{1,3} = 0$, $\Theta_1^{1,4} = 0.8$ and $\Theta_1^{1,5} = 1$. Then we conclude that offender group c_1^1 was active at three snapshots for any activity threshold equal to or less than 0.2. Statistics of this phenomenon is illustrated in Fig. 4.4. Even with small *activity thresholds* α, we do not have any offender group active in all time snapshots. With median α, we observe only a few offender groups which were active in three time snapshots. This may indicate that, due to reasons such as incarceration or changing crime-committing tactics and trends, offender groups generally do not maintain their co-offending activity for a long time period.

For considering a group active, we apply the activity threshold $\alpha = 0.3$. This means that a group is considered active if it maintains at least 30 % of its structure unchanged in the next time snapshot.

Offender Groups Size It is concluded [2] that most criminal organizations are quite small. Our study corroborates this result. Figure 4.5 provides the size distribution for known offender groups and active offender groups, and Fig. 4.6 shows the frequency of committed offences per group. Most groups committed less than 10 offences, but there are a few groups that committed even more than 100 offences during their life-cycle period.

Average group size for offender groups is 4.2 and for active offender groups is about 6.5. When comparing active offender groups to offender groups, a larger percentage of active offender groups has periphery members, and the average number of periphery members is greater, which may indicate that the periphery members play a more important role in the structure of active offender groups. The maximum number of kernel and periphery members in active offender groups, compared to offender groups, is significantly smaller. 7 % of the offender groups

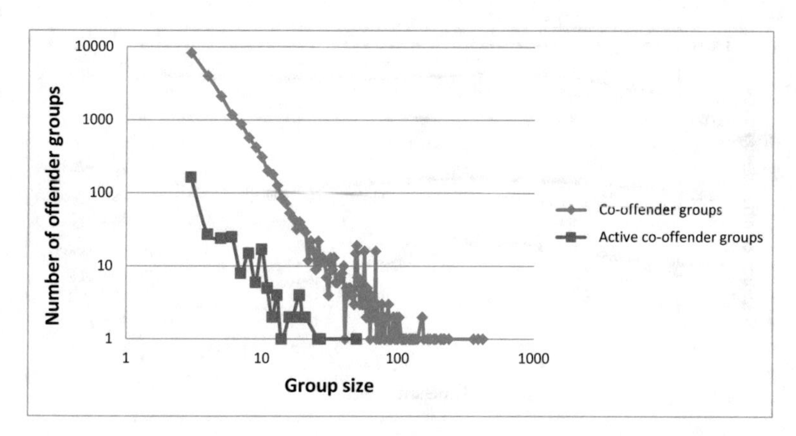

Fig. 4.5 Size of offender groups, active offender groups

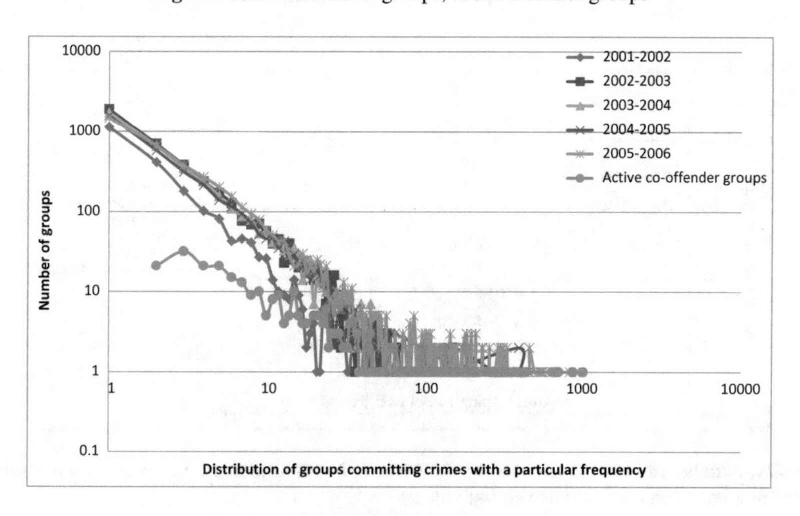

Fig. 4.6 Group committed crime frequency

have more than 10 and only 0.1 % have more than 50 members. In the active offender group set, 8 % of them have more than 10 members, and there is no group with more than 50 members. Of course, the size of these groups would likely be larger if all offences (not just known offences) were available.

Offender Groups Evolution Offender groups, similar to any other form of social community, typically evolve over time. An offender group may grow by admitting new members, shrink by losing members, split into two or more groups, or a new group may form by merging two or more existing groups. Given the limited observable time span, it is difficult to quantify the whole lifecycle of offender groups, not knowing their history previous to the first time step and their future history past the last time step. Figure 4.7 shows the statistics of different evolution scenarios in the five studied snapshots. For the matching function, the threshold value 0.3 applies for considering a group as survived (that is, it continues to exist in

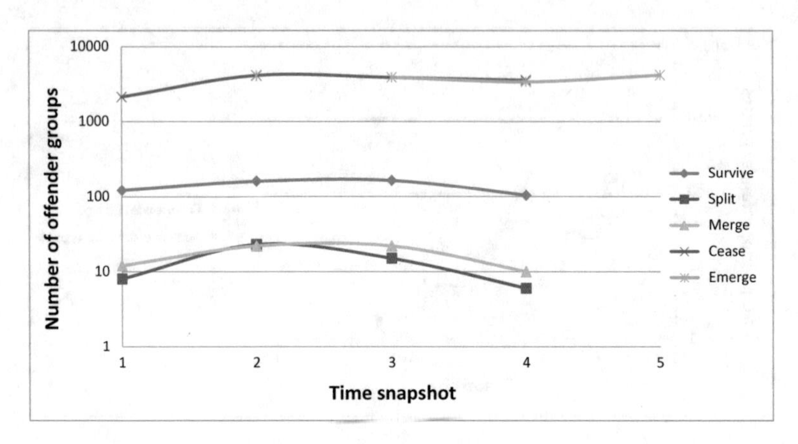

Fig. 4.7 Offender groups evolution trace

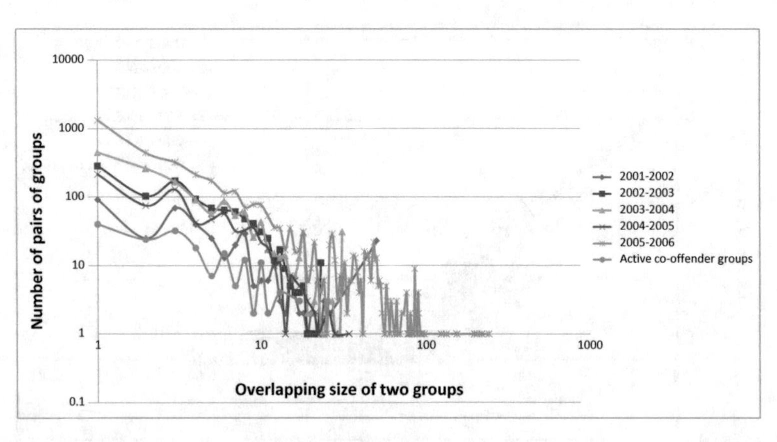

Fig. 4.8 Number of shared members for the overlapping co-offender and active offender groups by the number of pairs of overlapped individuals

any subsequent time snapshot) and a value greater than 0.2 and smaller than 0.3 for split and merge, respectively. Groups with matching thresholds smaller than 0.2 are considered as ceased groups (that is, not visible in any subsequent time snapshot). Over the 5 years of data, about 4 % of all offender groups survive, but split and merge events occur rarely, less than 1 % of the groups. About 96 % of the offender groups are considered ceased, since we do not observe their activity in the next time step, and 95 % of all groups are newly emerged ones.

Offender Groups Overlapping Figure 4.8 presents the distribution of size of overlap for offender groups and for active offender groups. For both group types the result is fairly similar. We see higher numbers for smaller sizes of overlap, which was predictable due to the applied method which is designed based on a strict definition of communities in the networks. Using a less strict definition of offender groups means that many of the currently overlapping groups merge

Table 4.1 Crime seriousness hierarchy and values (sample)

Crime type	Hierarchy level	Seriousness
Murder 1st degree	1	1
Abduction of person under 14	18	0.89
Production of heroin	41	0.74
Break and enter, residence	58	0.62
Theft of automobile	75	0.52
Theft over $5000—bicycles	83	0.46

into larger groups. In some cases, we observe several pairs of groups with more overlap. This is also because the applied method even differentiates between groups that have common periphery members but completely different kernel members. Regarding serious groups, there is little observable overlap, which again confirms their completely different structure compared to offender groups and active offender groups.

4.4.2 *Organized Crime Groups*

For calculating offender group criminality, we apply the RCMP crime seriousness index as delineated in the Operational Statistics Reporting System (OSR). This index uses a seriousness hierarchy with 151 groups, where each crime type belongs to one of these groups. For each crime type in the dataset, the corresponding seriousness group level is scaled linearly, and these normalized values are interpreted as indicators of the seriousness of offences. Table 4.1 shows a small sample of the OSR crime seriousness hierarchy and corresponding seriousness values.

Figures 4.9 and 4.10 illustrate the number of offender groups in respect of different criminality thresholds β, where β is equal to the summation of seriousness of offences committed by members of a co-offending group divided by the total number of those offences. About 30 % of all offender groups pass the threshold $\beta = 0.6$, which means a larger percentage of the offender groups commit minor crimes, which is intuitive. Finally, $\beta = 0.8$ identifies less than 6 % of the groups, which implies that a small percentage of offender groups are consistently involved in serious crimes.

Finally, the possible organized crime groups are extracted from active offender groups. Applying activity threshold 0.3, a total of 313 groups are considered active offender groups. Figure 4.10 shows the number of active offender groups for different criminality thresholds. From 313 active offender groups 89, 39, 18, 8, and 5 groups result for criminality thresholds 0.5, 0.6, 0.7, 0.8, and 0.9, respectively. There is no active offender group having criminality equal to 1.

An important question is about the applied criminality thresholds for detecting possible organized crime groups. For this purpose we use the average seriousness

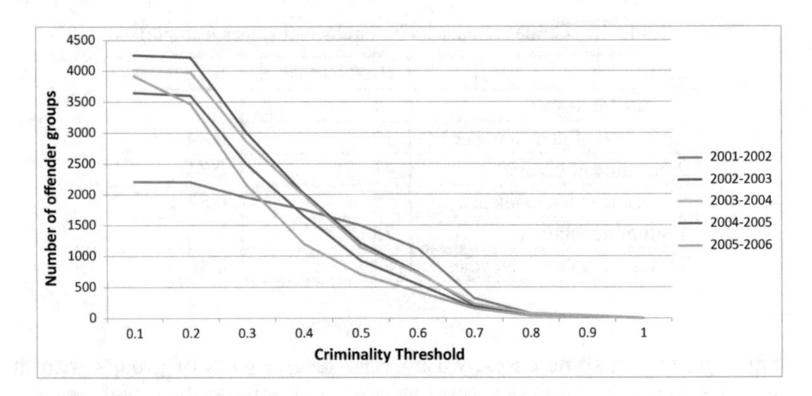

Fig. 4.9 Number of offender groups in respect of different criminality thresholds

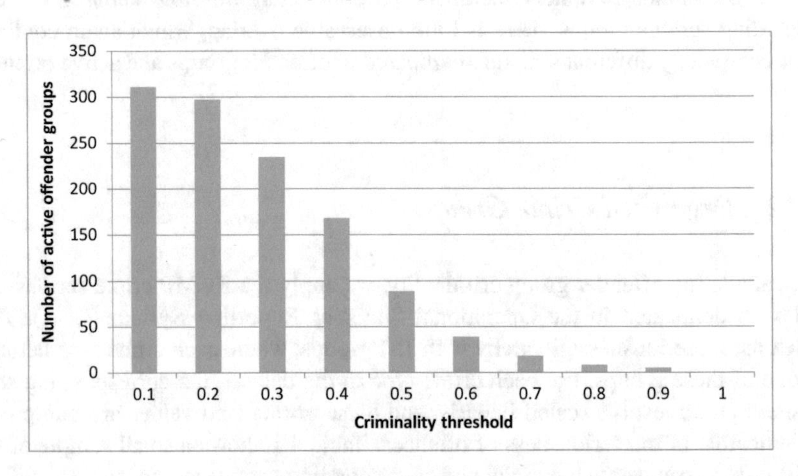

Fig. 4.10 Number of active offender groups in respect of different criminality thresholds

of all serious offences (defined in the PS serious crimes list), which is 0.6. In total, 39 groups have criminality higher than 0.6 which are considered possible organized crime groups.

As an alternative approach for computing criminality of groups, we also test the use of the crimes' seriousness as defined in the Crime Severity Index (CSI) list. The weights used in the CSI were developed by the Canadian Centre for Justice Statistics, Statistics Canada and the Canadian law enforcement community to empirically compare the relative seriousness of criminal offending. In the CSI offences are weighted by the relative seriousness by which they are treated by Canadian courts. Our experiments find that applying the CSI to analyze group criminalities using our present method results in findings that are much skewed and we do not see a distributed group criminality as for the other approach (OSR crimes seriousness list). The reason for this is that in the OSR list (mapped to the OSR list) crimes are assigned a seriousness value between 1 and 151. But in the CSI list the maximum and minimum values are 7041.45 and 1.16, and 90 % of the crimes seriousness has a value less than 500. Therefore the group criminality results in a skewed-left distribution, with very few active offender groups identified as serious offender groups. Future research will need to develop an efficient approach to normalize crimes seriousness to get more distributed group criminality.

In the 5 years of the crime data, using the soft constraint approach, there are 39 groups that are both active and serious and thus are considered possible organized crime groups. Interestingly, most of these groups have high activity, which shows the close relationship among their group members. The average size in this set of groups is 4.7, which is much smaller than the size of active offender groups. This point supports the theory that with increasing group criminality offender group size decreases. Having less number of possible organized crime groups with periphery members compared to active offender groups also implies that in possible organized crime groups, the kernel members are not eager to collaborate with co-offenders outside of the group's core.

Hard Constraint Approach The "hard constraint approach" is the most basic test to determine possible organized crime groups in the dataset since the approach most closely matches the Criminal Code definition of criminal organization [9]. In the hard constraint approach all crimes are categorized into a binary classification of the two classes: (1) serious crimes with material benefit; and (2) non-serious crimes or serious crimes that do not appear to provide the offender group with a material benefit.

We use the list of serious offences prepared and provided by Public Safety Canada (PS) officials. The list was developed as follows [15]:

- First, the list of all offences in the Criminal Code was taken as the preliminary set. The criteria for seriousness—offences subject to indictment and having a penalty of five or more years imprisonment—was then applied to the full set of Criminal Code offences, with all offences not meeting this requirement removed from the set. As some offences (commonly referred to as hybrid offences) may proceed either by indictment or by summary process, all offences where indictment was an option were included as in the list of section 467.1 "serious crimes."
- The set of offences this decision rule generated was then further examined to extract those offences with a "material benefit." Under section 467.1 and current case law "material benefit" could include an act that results in an intangible gain such as to one's criminal reputation as well as financial gain. The law on point is still evolving so the current list of material benefit crimes represents an approximation of material benefit offences based on examination of literature and other information related to organized criminality in Europe and the USA. Essentially, the list includes offences for which at least one known case of a direct material benefit (e.g., cash payment) or intangible benefit (e.g., increases one's criminal reputation) exists in either of the two named jurisdictions.

The BC crime dataset used in this study which is retrieved from the RCMP's Police Information Retrieval System (PIRS) identifies a very large set of offence category types. The PIRS list is a classification of calls for police services and contains information on offences and violations of many federal and provincial statutes as well as violations of the Criminal Code. The PIRS criminal event categories match the PS list of serious material benefit crimes to a substantial but incomplete extent: 112 out of 192 PS categories have a matching PIRS crime category. PS-identified offences that lacked a PIRS category match could not be used in this study because we could not attribute any of the offences under study to those PS categories. All PIRS offences that do not qualify as a serious offence resulting in a material benefit are disregarded in our analysis of the crimes committed by an offender group.

It should be noted that this study tests the feasibility of a new analysis technique for exploring networks in criminal organization using a historical database. Future work could use current data rather than historic data. With current data it would be feasible to determine how police crime categories are translated into Criminal Code sections in the process of laying charges and trying cases by linking police records with court records. This is not possible at this time in British Columbia with historic data. Future research could also make use of future modifications in Public Safety Canada classifications or in Canadian Centre for Justice Statistics (CCJS) data collection methods. As this study shows, there is value in improving analysis techniques concurrent with advancements in data collection and changes in information classification schemes.

The analysis presented in this section uses a binary classification of offence types according to whether or not an offence constitutes a serious offence that results in potential material benefit for an offender group committing the offence. All offences that do not qualify as a serious offence resulting in material benefit are disregarded in the analysis of the crimes committed by an offender group. This analysis used a binary classification scheme that used the categories of offences in the historic PIRS data and divided these offences as closely as possible to a list of serious crimes resulting in material benefits used by Public Safety Canada [15]. The Public Safety Canada list is crimes defined under relevant Criminal Codes which correspond directly with offences prosecuted in Canadian courts. The PIRS list is a classification of calls for police services and contains some categories that are not provided in detail under the Criminal Code. As well the PIRS categories do not contain all of the categories in the Public Safety Canada list. But the classification used does divide police classification by seriousness resulting in material benefit.

The following presentation and discussion of results of the analysis is strictly based on this binary classification of the hard constraint approach. We can identify serious groups based on two aspects: (1) the ratio of committed serious offences to total committed crimes, and (2) the number of committed serious offences. In the first approach, an offender group is considered serious if $P\,\%$ of all offences committed by this group is serious in the above sense. In the second approach, we consider a group serious if two or more members of this group were involved in more than N serious offences (where $P\,\%$ and$N$ refer to adjustable threshold values for the percentage and number of serious offences committed by a group, introduced for the purpose of controlling the analysis). Further to the overall approach taken here, another possible way would be to calculate the ratio of serious crimes to the number of individuals in the group.

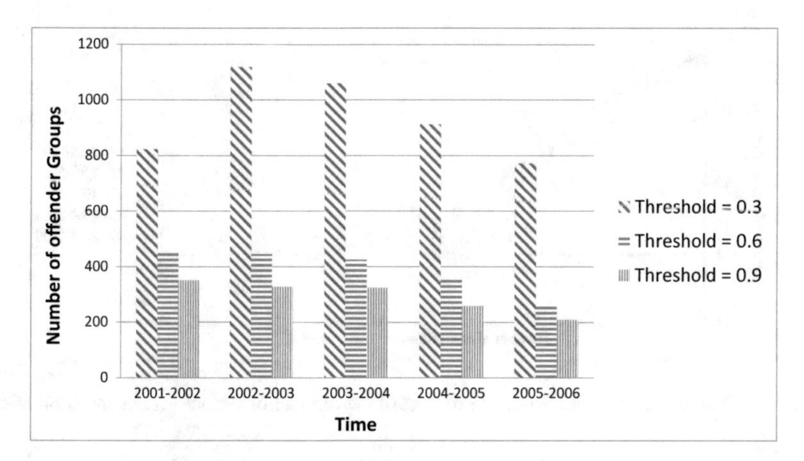

Fig. 4.11 Number of observed offender groups in respect of the proportion of committed serious crimes

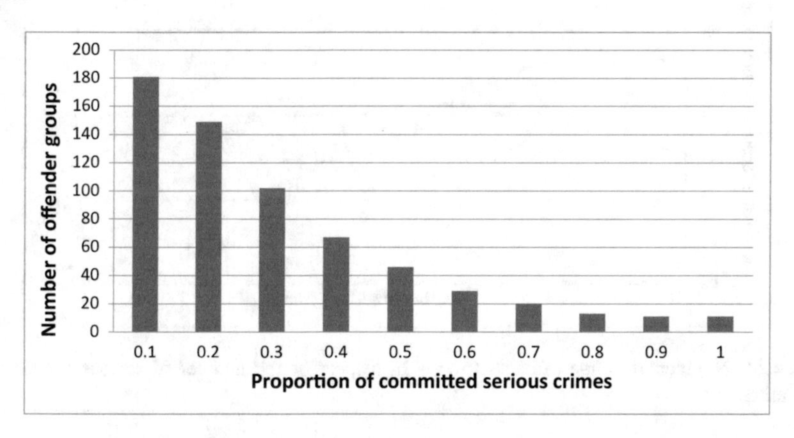

Fig. 4.12 Number of observed active offender groups in respect of the proportion of committed serious crimes

In the hard constraint approach to detecting possible organized crime groups, we consider the ratio of a group's committed serious crimes to total committed crimes. Figures 4.11 and 4.12, respectively, show the number of offender groups and active offender groups in respect of the percentage of committed serious crimes. Considering different percentage thresholds, $P = 30$, $P = 60$, and $P = 90$, respectively, 25, 10, and 8 % of the offender groups and 33, 9, and 4 % of the active offender groups remain in the list.

Figures 4.13 and 4.14, respectively, show the number of offender groups and active offender groups in respect of the number of serious crimes these groups have committed over their life cycle. In the offender group set, about 59 % did not commit any serious crime; 91 % of them were involved in less than five serious crimes and only 0.02 % of the offender groups committed more than 10 serious crimes.

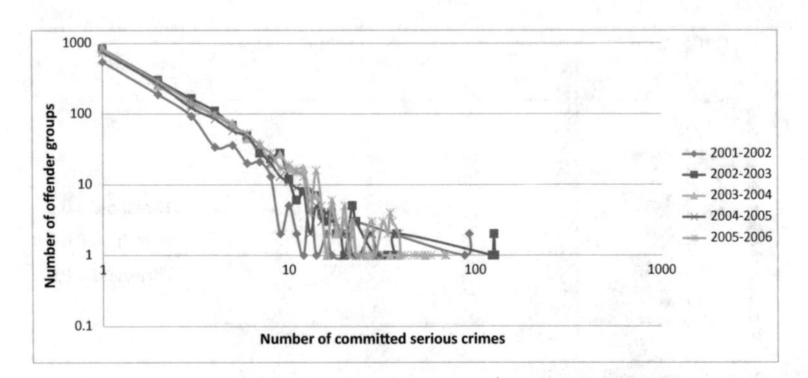

Fig. 4.13 Number of offender groups in respect of the number of serious crimes they committed

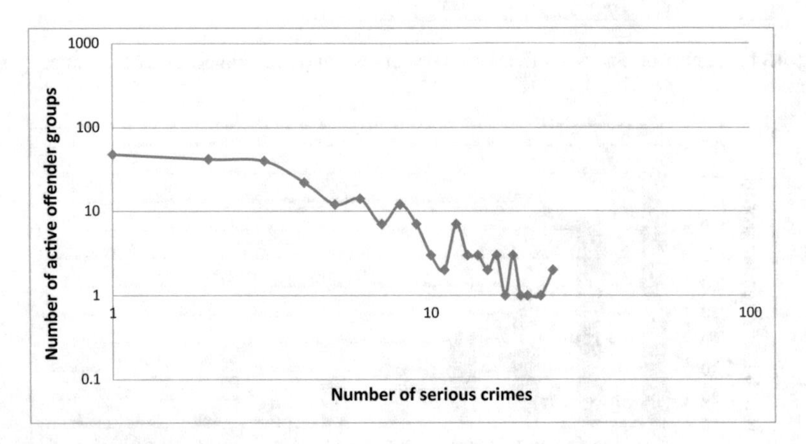

Fig. 4.14 Number of active offender groups in respect of the number of serious crimes they committed

The average number of serious crimes per offender group is 1.2. In total, 25 % of the active offender groups did not commit any serious crime, while 73 % of these groups committed less than five serious crimes during their life cycle and only 0.09 % of the active groups committed more than 10 serious crimes. On average, each active offender group has committed 3.7 serious offences. These results show that, compared to the offender groups, active offender groups more frequently commit serious offences.

In the hard constraint approach, based on the definition of organized crime in the *Criminal Code*, we consider any active offender group that committed at least one serious offence. In this case, out of 313 active offender groups, we end up with a list of 236 groups. These 236 groups meet the minimum analytic threshold to be considered possible organized crime groups. In total, 49 of these groups are involved in only one serious offence and one of them committed 24 serious offences.

4.5 Conclusions

Controlling crime necessitates the investigation of criminal networks, criminal organizations and their illegal activities, constituting a serious undertaking for law enforcement and the criminal justice system. We propose here a computational co-offending network analysis approach for detecting possible organized crime groups. We evaluate the proposed methods by examination of a large real-world crime dataset. Our examination shows that although criminal group activity does not occur as routinely as other criminal activities, which is intuitive, there is continuous criminal collaboration inside crime groups. But for most of the groups such co-offending behaviour does not persist over longer time periods. Our study also shows that active offender groups typically have more peripheral members in contrast to serious groups which tend to have fewer peripheral members and a tightly connected kernel. This finding suggests that serious groups operate primarily from inside their core membership.

Starting from a crime dataset with 4.4 million records and a co-offending network with 150,000 actors, we were able to detect more than 18,000 offender groups, including more than 300 active groups and 39 possible organized crime groups. Using the hard constraint approach, our study identifies 236 possible organized crime groups which committed one or more than one serious offences over the observed time frame.

Our analytic approach provides potentially important insights into the ways in which co-offending networks shape and affect criminal behaviour. Albeit, it should be noted that co-offending networks do not necessarily identify all individuals of an organization, simply because those operating in the background, who often direct the activities of others, may not be visible in the data. Another possibility is that some of the detected organized crime groups represent particular functional components of larger criminal organizations that do not appear directly in police-reported crime. For obtaining a more holistic picture of criminal organizations, one must combine police-reported crime data with data from intelligence agencies. This type of analysis can identify possible individuals or offenders for further investigation or as part of a disruption strategy. Further, the approach taken here primarily concentrates on organized crime groups with dense member relationships, which is not always the case, especially not for certain forms of criminal networks.

A major advance could flow from undertaking research with current police data that is linked to current court data. This could make it possible to use a definition firmly based on the Criminal Code by developing a probabilistic concordance of crimes with police categories of calls for police service. It would also be possible to merge datasets of associations between persons that are not based on co-offending.

References

1. J. Abello, M.G.C. Resende, S. Sudarsky, Massive quasi-clique detection. in *LATIN 2002: Theoretical Informatics* (Springer, Berlin, 2002), pp. 598–612
2. J.S. Albanese, North American organised crime. Glob. Crime **6**(1), 8–18 (2004)
3. A.A. Block, *East Side-West Side: Organizing Crime in New York City, 1930–1950* (Transaction Publishers, New Brunswick, 1994)
4. S.P. Borgatti, M.G. Everett, P.R. Shirey, LS sets, lambda sets and other cohesive subsets. Soc. Networks **12**(4), 337–357 (1990)
5. U. Brandes, D. Delling, M. Gaertler, R. Gorke, M. Hoefer, Z. Nikoloski, D. Wagner, On modularity clustering. IEEE Trans. Knowl. Data Eng. **20**(2), 172–188 (2008)
6. M. Carlo, *Inside Criminal Networks* (Springer, Berlin, 2009)
7. D. Chakrabarti, R. Kumar, A. Tomkins, Evolutionary clustering, in *Proceedings of the 12th ACM SIGKDD International Conference on Knowledge Discovery and Data Mining (KDD'06)*, pp. 554–560, 2006
8. C. Fijnaut, F. Bovenkerk, G. Bruinsma, H. van de Bunt, *Organized Crime in the Netherlands* (Kluwer Law International, The Hague, 1998)
9. U. Glässer, M.A. Taybei, P.L. Brantingham, P.J. Brantingham, Estimating possible criminal organizations from co-offending data. Public Safety Canada, 2012
10. International report on crime prevention and community safety: Trends and perspectives, 2010. Retrieved from http://www.crime-prevention-intl.org/fileadmin/user_upload/Publications/Crime_Prevention_and_Community_Safety_ANG.pdf, April 2010
11. Y. Lin, Y. Chi, S. Zhu, H. Sundaram, B.L. Tseng, Facetnet: a framework for analyzing communities and their evolutions in dynamic networks, in *Proceedings of the 17th international conference on World Wide Web (WWW'08)* (ACM, New York, 2008), pp. 685–694
12. D.W. McMillan, D.M. Chavis, Sense of community: a definition and theory. J. Community Psychol. **14**(1), 6–23 (1986)
13. M.E.J. Newman, Fast algorithm for detecting community structure in networks. Phys. Rev. E **69**(6), 066133 (2004)
14. G. Palla, I. Derényi, I. Farkas, T. Vicsek, Uncovering the overlapping community structure of complex networks in nature and society. Nature **435**(7043), 814–818 (2005)
15. M. Saunders, Personal communication with public safety Canada. Public Safety Canada,May, 2012
16. M. Spiliopoulou, I. Ntoutsi, Y. Theodoridis, R. Schult, Monic: modeling and monitoring cluster transitions. in *Proceedings of the 12th ACM SIGKDD International Conference on Knowledge Discovery and Data Mining (KDD'06)*, pp. 706–711, 2006
17. M.A. Tayebi, U. Glässer, Investigating organized crime groups: a social network analysis perspective, in *Proceedings of the 2012 International Conference on Advances in Social Networks Analysis and Mining (ASONAM'12)*, pp. 565–572, 2012
18. T. van der Heijden, Measuring organized crime in Western Europe. Retrieved from https://www.ncjrs.gov/policing/mea313.htm, 1996
19. K. von Lampe, Organized crime in Europe: conceptions and realities. Policing J. Policy Pract. **2**(1), 7–17 (2008)
20. K. von Lampe, Definitions of organized crime. Retrieved from http://www.organized-crime.de/organizedcrimedefinitions.htm, February 2015
21. K. Yu, S. Yu, V. Tresp, Soft clustering on graphs, in *Advances in Neural Information Processing Systems (NIPS'05)*, pp. 1553–1560, 2005

Chapter 5
Suspects Investigation

In their efforts to identify potential suspects, crime investigators routinely draw on partial knowledge as the result of incomplete information and uncertain clues. Physical evidence gathered at a crime scene as well as accounts from victims and witnesses may be incomplete and inconclusive. In cases with multiple offenders jointly committing a crime, where individual suspects have been identified, the aim of co-offending network analysis is to complement *criminal profiling methods* [3, 6] so as to identify additional suspects faster and more effectively, thus decreasing the cost and time of crime investigations.

A common example is organized crime [4] as a form of criminal activity following a regular pattern, such as continuity or other spatiotemporal characteristics in contrast to irregular criminal behaviour associated with opportunistic crime. Law enforcement agencies often gain partial information about organized crime structures, for instance, from arrested suspects and convicted offenders who confess their affiliation with an organized crime group. However, uncovering the whole structure of criminal organizations under investigation poses a considerable challenge for law enforcement. Systematic approaches to co-offending network analysis can help tremendously in such cases.

While many research studies in the literature use co-offending networks for crime suspect investigation, for instance [18, 19], to the best of our knowledge, none of these works define this problem formally by proposing an algorithmic solution.

In this chapter, we introduce the problem of crime suspect recommendation as the goal to *recommend the top-N additional potential suspects*, given a partial set of crime suspects and a known co-offending network. To address this problem, we propose a random walk based method, called CRIMEWALKER, for link prediction and scenarios with a set of given suspects instead of a single source user.

Section 5.1 explores related work, and Sect. 5.2 introduces a formal definition of the suspect investigation problem. Section 5.3 then presents CRIMEWALKER, while Sect. 5.4 describes the results of our experimental evaluation. Section 5.5 concludes this chapter.

M.A. Tayebi, U. Glässer, *Social Network Analysis in Predictive Policing*,
Lecture Notes in Social Networks, DOI 10.1007/978-3-319-41492-8_5

5.1 Background

Generally, crime analysis captures a broad spectrum of facets pertaining to different needs and using different analytical methods, namely: administrative analysis, strategic analysis, tactical analysis, criminal investigative analysis, and intelligence analysis.

Administrative crime analysis aims at reducing or preventing crime by reporting local and regional statistics of crime rates to higher-ranking managers of law enforcement agencies. Strategic crime analysis primarily focuses on planning strategies for crime reduction and prevention. Tactical analysis tries to recognize repeating crime patterns and serial crimes and identify potential suspects for such crimes; it also makes predictions of future crime based on extracted patterns. Criminal investigation analysis focuses on collecting, testing, and keeping information of known offenders in serious crimes. Finally, intelligence analysis aims at recognizing relationships between criminal network actors in order to identify and arrest these offenders. Intelligence analysis typically starts with a known crime problem or identified co-offending network and then uses these resources to collect, analyze, and compile information about a predetermined target [5].

Crime investigation serves the purpose of discovering the sole truth or, in case of many potential possibilities, narrowing down the scope to the closest possibility. An essential task in intelligence analysis is *criminal profiling*, the process of using crime scene evidence to reconstruct missing information and reason about potential suspects, including personality characteristics and psychopathology. In traditional criminal profiling, information is gleaned from the crime scene and linked to the behaviour of an offender during a crime. The source of such information can be accounts by victims or witnesses and includes age range, gender, home location, and other characteristics of an offender, which can then be used to aid police in their investigations. More recently, formal reasoning approaches in support of crime investigation have been studied [3, 6]. Specifically methods like Bayesian networks and neural network are used for criminal profiling, which can lead investigators to potential suspects.

Law enforcement and intelligence agencies have realized that detailed knowledge about criminal networks can be invaluable for crime investigations [15]. PFS [19] is a link analysis technique using shortest-path algorithms to identify the strongest association paths among criminal network entities. In evaluation studies domain experts conclude that the association paths identified by PFS algorithms are helpful in 70 % of the cases. Xia [18] proposes MPFS, an extended version of PFS [19]. His approach uses belief propagation to update the belief of crime investigators as to the probability for an individual member of a criminal network being an offender. Building on existing partial information, this method is able to choose the next member of a given network as a candidate for further investigation.

Since the work presented here builds on social recommendation methods for suspect investigation, we also review some related work on recommendation in social networks.

Authors of [11] propose a random walk method (TrustWalker) which combines trust-based and item-based recommendation. TrustWalker considers not only ratings of the target item, but also those of similar items. The probability of using the rating of a similar item instead of a rating for the target item increases with increasing length of the walk. This framework contains both trust-based and item-based collaborative filtering recommendations as special cases. The experiments show that their method outperforms existing memory based approaches. We extend this method to address the suspect investigation problem in this chapter. Note that the input to TrustWalker is a single user, while the input to our problem is a set of offenders already charged with the crime.

SocialMF [12] is a matrix factorization based approach for recommendation in social networks. In SocialMF, the rating matrix is decomposed into a product of latent feature vectors of users and items. To take the social network into account, SocialMF considers the latent feature of a user to be close to the aggregate of the latent features of his direct neighbors. SocialMF also supports trust propagation between indirect neighbors in the social network.

5.2 Problem Definition

For the new crime event e, we assume that police charges a subset of offenders A linked to the crime e. Then, police want to investigate additional suspects potentially involved in e based on the existing information, that is, the known offenders A who committed e and the relations between all offenders in crime dataset $\mathscr{C}$ as represented by co-offending network G. The problem can now be specified in abstract formal terms as follows [17]:

> Given a crime dataset $\mathscr{C}$, a co-offending network $G(V,E)$ defined on $\mathscr{C}$, and a new crime event e with a set A of offenders already charged, recommend the top-N suspects not included in A that are co-offenders in the crime event e with high probability.

This problem definition is novel, especially with its application in the domain of criminal networks.

5.3 CrimeWalker

This section introduces, CrimeWalker [17], our random walk based model for recommending top-N suspects for a crime event. Random walk based models for recommendation in social networks have been investigated recently [8, 11]. These methods exploit the random walk with restart (RWR) method for recommendation in a social rating network. A social rating network is a social network in which users are allowed to express ratings on some items besides being able to create social relations to other users.

In RWR, the goal is to recommend top-N nodes to a given node i. A random walker starts to walk along the edges of the graph. At each node j, the random walker stops and restarts the walk from i by a fixed probability c. The similarity score of node j with node i is defined as the probability $S(i,j)$ that the walker will reach j where this probability is computed iteratively as follows:

$$\mathbf{S_i} = (1-c)P\mathbf{S_i} + c\mathbf{e_i} \tag{5.1}$$

P is the $N \times N$ transition matrix of graph G. $\mathbf{e_i}$ is $N \times 1$ starting vector that the ath element is 1 and the others is 0, and $\mathbf{S_i}$ is an $N \times 1$ vector that the j element of this vector shows the similarity of node i to node j. TrustWalker [11] extends RWR to address the problem of link prediction. TrustWalker considers the stopping probability c to be dependent on the rating pattern similarity of users. Also the probability of random walk to restart increases with the increasing length of a walk in TrustWalker. Due to the complexity of the model in TrustWalker, it actually performs the random walks to approximate the probabilities of reaching different nodes. We extend TrustWalker and propose CRIMEWALKER to address the suspect investigation problem. Given graph G and offender u, CRIMEWALKER performs a set of random walks on the co-offending network to compute the recommended offenders suspicious of co-offending with u. For some offenders that random walks do not converge we terminate the whole process after a fixed number of random walks. Unlike TrustWalker, each random walk in CRIMEWALKER returns a node in the co-offending network (a potential co-offender). The results of a set of random walks are aggregated together to compute a list of top-N co-offenders.

Note that the input to CRIMEWALKER is a crime event e with a set A of already charged offenders instead of a single offender. Therefore CRIMEWALKER has to combine the results of recommended co-offenders for all the offenders in the input data A into one single top-N result. We first discuss the details of a single random walk in CRIMEWALKER. Then we present our proposed method to combine the results of different input offenders.

5.3.1 A Single Random Walk in CRIMEWALKER

In CRIMEWALKER, a set of random walks are performed for a single offender u to compute the top-N suspects recommended to be co-offending with u. Every single random walk starts from u and continues walking on the co-offending network. When the random walk stops, the offender node at which it stops is returned as the result of this random walk. There are two probabilities that should be discussed in the random walk. First the probability of stopping at a node v:

$$\phi_{v,u,k} = \mathrm{Sim}(u,v) \times \frac{1}{1+e^{-k/2}} \tag{5.2}$$

Here, $\phi_{v,u,k}$ is the probability of stopping at node v at the kth step of a random walk starting from node u. $\mathrm{Sim}(u,v)$ is the similarity measure introduced in Sect. 3.3. Similar to TrustWalker, we punish long random walks by taking the length of the random walk into account. Second, the probability $p(w|u,v,k)$ of walking from node v to node w in the kth step of the random walk starting from u is computed as follows:

$$p(w|u,v,k) = \frac{n_{v,w}}{\sum\limits_{y \in \Gamma_v} n_{v,y}} \tag{5.3}$$

where $n_{v,w}$ is the strength (refer to Sect. 3.1.2) of the edge between v and w, which is equal to the number of crimes they have co-offended. Also, Γ_v is the set of neighbors of v in the co-offending network.

After performing a set of random walks, top-N suspects with the highest frequency of being returned as the result of a random walk are considered as the top-N recommended co-offenders for the input offender u. Note that CrimeWalker approximates the probability $p(v|u)$, the probability of reaching v in a random walk starting from node u by the number of times v is returned as the result of the random walk. In the following subsection, we propose our approach to combine the results of different input offenders into a single top-N recommended suspects.

5.3.2 CrimeWalker *for a Set of Offenders*

Since we have a set of offenders as the input, we have to combine the co-offenders recommended for every single offender to have a final set of recommended offenders. For every offender a in the set A of already charged offenders, we compute a set R_a of recommended suspects. Every recommended suspect $r_a \in R_a$, is associated with a probability $p(r_a|a)$ indicating the probability of reaching r_a in a random walk starting from node a. There are two ways to combine the recommended set of suspect. In the first approach, the probability of each suspect being reached in random walks starting from different charged offenders is aggregated to have an aggregate probability:

$$\forall u \in V, p(v|A) = \frac{\sum\limits_{a \in A} p(v|a)}{|A|} \tag{5.4}$$

In other words, we can introduce a super node S_A that has direct links to all nodes in A (with equal weight), and perform a random walk based approach for the super node as a single node. Another approach to combine the lists of recommended suspects is to be more tough and assign high probability to suspects that are recommended for more charged offenders. Basically, the probability of reaching every node from the charged offenders in A is multiplied to compute an aggregate probability:

$$\forall u \in V, p(v|A) = \prod_{a \in A} p(v|a) \tag{5.5}$$

These two approaches have two different conceptual meanings: the CRIMEWALKER-Disj method favors suspects that are on average suspicious according to the given set of offenders. This approach does not distinguish between the following cases: first, a suspect is highly recommended to be co-offender of a charged offender while not very likely to be co-offender of another charge offender. Second, a suspect is moderately suspicious of co-offending with both charged co-offenders. On average, both suspects are equally likely to be the co-offender and CRIMEWALKER-Disj considers them to be equally suspicious of being the co-offender. On the other hand, the CRIMEWALKER-Conj approach punishes the cases where the suspect is not very likely to be co-offending with one of the charged offenders. In other words, CRIMEWALKER-Conj favors the second sample case more than the first case.

5.3.3 *Similarity Measure for Offenders*

We consider two common factors directly affecting the probability of two offenders committing crimes together: How strong their relationship is, and how similar they are regarding personal characteristics, including age, education, ethnic group, gender, personal contacts, social background, etc.

The homophily principle states that individuals in a social network tend to be similar to their friends [13, 16], specifically they often interact with others similar to themselves. There are two driving factors: *selection* and *social influence*. According to selection, people tend to create social relations to other people who have similar personal attributes. Social influence leads to people adopting the attributes of their friends in a social network. For example, Vancouverites are more likely to be connected to other Vancouverites rather than to individuals of another city. People usually choose the least distance necessary to get what they need, so their social interactions tend to happen in close distance to their home; thus, one can expect a higher probability for a relationship between two offenders if they live close to each other.

To take the attributes of the offenders into account we define a similarity measure based on the characteristics of the offenders including their age, sex, and living location. Similarity of offenders u and v, $\mathrm{Sim}(u,v)$ is computed as follows:

$$\mathrm{Sim}(u,v) = 1 - \left(w_d \times \frac{D_{uv}}{\mathrm{Max}(D)} + w_a \times \frac{A_{uv}}{\mathrm{Max}(A)} + w_g \times G_{uv} \right) \tag{5.6}$$

where D_{uv} and $Max(D)$, respectively, denote the distance between the home locations of u and v and the maximum distance among all pairs of offenders. A_{uv} and $Max(A)$, respectively, denote the age difference of u and v and the maximum age difference between all offenders. And G_{uv} is equal to 0, if offenders u and v have the same gender, and 1, otherwise. w_d, w_a, and w_g, respectively, denote the weight assigned to each of these three factors to balance their importance. In the next section we describe how we compute the feature weights.

5.3.4 *Feature Weights Computation*

We use χ^2 (chi-square) method to define the weight of the features. χ^2 test in statistics is used to test the independence of two events. The χ^2 aggregates the deviation of observed values from expected values, under the independence hypothesis. For our case the independence hypothesis states that a feature plays no role in forming a co-offending link. Pair of offenders (u,v) is in connected class if they are connected in the co-offending network, or they are in non-connected class if they are not connected. We associate a binary value for every feature of offenders u and v. If they match on a feature this value is 1, otherwise it is 0. For multi-value features such as home location distance a threshold function defines if u and v matches or not. In other words, if home location distance of these offenders is smaller than the predefined threshold they match, and vice versa. The χ^2 measure is computed as follows:

$$\chi^2 = \sum_i \frac{(O_i - E_i)^2}{E_i} \tag{5.7}$$

where O_i is the observed frequency of samples in category i, and E_i is the expected frequency of samples in category i. For each of the features we compute the χ^2 score, and eventually use the corresponding normalized value as weight of that feature.

5.4 Experiments and Results

In this section, we perform experiments on different variations of CRIMEWALKER and the comparison partners to compare the effectiveness of each approach [17].

5.4.1 *Experimental Design*

We divide the BC crime dataset chronologically into 90 % train and 10 % test data. For each crime event in the test data, we withhold an offender (one by one for every offender in the crime event) and ask the recommendation model to recommend

Table 5.1 Statistical properties of the DBLP co-authorship network

Metric	Value
Nodes of authors	663,620
Average degree	6.9
Average distance	3.64
Diameter	9
Effective diameter	4.49
Average clustering coefficient	0.65
Largest component percentage	0.85

top-N suspects who may have collaborated in this crime. N is set to 10 in our experiments. The relative results for other values of K are consistent.

DBLP Dataset The crime data used are very sensitive and it is not available for public, therefore, there is no chance to repeat the experiments, and we use the DBLP bibliography [1] dataset as well. This dataset is an event based dataset similar to the crime data. Using this dataset we generate a co-authorship network. For this purpose papers which appeared in a 11-year period (1999–2009) are used. This data is converted into a weighted co-authorship network, where each author is denoted as a node, an edge between two authors shows a joint publication between these two authors and weight of an edge between two authors represents the frequency of their co-authorships. The network spanning 11 years contains 663,620 nodes and 4562876 edges. Statistical properties of the co-authorship network extracted from DBLP data are presented in Table 5.1.

In the DBLP data set, we use papers published in 2009 with authors who have had at least one paper in the period (1999–2008) as the test data, which includes 64,723 papers. We use the recall measure of the withheld offenders to evaluate the quality of the recommendation model. Recall is the percentage of recommendation queries in which the withheld user appears in the list of top-N recommended users.

5.4.2 Comparison Partners

We evaluate and compare the result of the CRIMEWALKER-Disj and CRIMEWALKER-Conj method introduced in Sect. 5.3.2 on both crime data and the DBLP dataset. To evaluate the meaningfulness of co-offending links strength, and also the similarity of offenders, we perform experiments on different settings of our proposed method. Following is the list of all the comparison partners in our experiments.

[1] http://dblp.uni.rier.de/.

CrimeWalker-DNW This setting of CrimeWalker-Disj ignores the weights of edges in the random walk and also the similarity of nodes in stopping criteria. Note that this method is the closest comparison partner to the RWR method. However, unlike RWR, it has the advantage of punishing long walks.

CrimeWalker-DNS This version of CrimeWalker-Disj takes the links strength into account but ignores the similarity of nodes.

CrimeWalker-Disj This is the full CrimeWalker-Disj.

CrimeWalker-CNW This setting of CrimeWalker-Conj ignores the weights of edges in the random walk and also the similarity of nodes in stopping criteria.

CrimeWalker-CNS This version of CrimeWalker-Conj takes the links strength into account but ignores the similarity of nodes.

CrimeWalker-Conj This is the full CrimeWalker-Conj.

AR This method is an association rule mining approach and is used as the baseline in our experiments. In the following we briefly discuss the details of how we exploit association rule mining in our problem.

Agrawal et al. [1] introduced the framework of association rules into the data mining community. Representational aspect of association rules had been discussed by Hajek et al. [9], but their focus was not on the algorithmic aspects of rule mining. Different association rule mining approaches have been published in the data mining literature such as Apriori [2] and FP-growth [10].

We now introduce the basic terminology of association rule mining. A transaction is a set of items. An association rule is a rule in the form of $A \Rightarrow B$, where A and B are sets of items. This rule is supposed to mean that the presence of A in a transaction implies the presence of B with some probability in that transaction. Each rule has two relevant measures for a set of transactions: support and confidence. Support is the percentage of transactions that include both A and B in all transactions. Confidence is the percentage of transactions including B in the set of transactions that include A.

Association rules detect relationships between items considering co-occurrence patterns in the transactions set. This method can be an appropriate approach for recommendation where there exist transaction data and has been used widely in this regard [7, 14]. Crime data can be interpreted as transaction data. In this case, a transaction is a set of offenders involved in a crime and each item corresponds to an offender.

Assume that we have a set of A of offenders involved in a crime event in the test data, and we withhold offender u from this set. The rule $A - u \Rightarrow u$ is capable of recommending the withheld offender u. The rule mining method is interested in subset of association rules whose heads are restricted to the withheld suspect offender and the rule's body is the other involved offenders in that crime. Moreover, we also take any rule with the body consisting of any subset of $A - u$ into account. Meaning that, not only we consider rules with all the charged offenders in its body, but also we consider the rules having a subset of the charged offenders in their body. The rule mining recommendation method is looking for the strongest rules.

Table 5.2 CrimeWalker results on the BC co-offending network

Method	Recall (%)
CrimeWalker-DNW	6.6
CrimeWalker-DNS	7
CrimeWalker-Disj	7.5
CrimeWalker-CNW	7.5
CrimeWalker-CNS	7.8
CrimeWalker-Conj	8.4
AR	5.0

For each target offender we choose the ten strongest rules based on the support and confidence measures. If any of these ten rules suggest the target offender correctly, the recall measure increases.

5.4.3 Experiments and Results

Table 5.2 shows the experimental results for the comparison partners on the BC crime dataset. As shown in this table, random walk based methods clearly outperform the association rule mining based method. We believe that this is mainly due to the fact that random walk based methods consider the transitivity effect and take the co-offenders of an offender into account for identifying potential suspects.

According to the results of CrimeWalker on the BC crime dataset, exploiting the links strength and similarity measures improves the quality of recommendation. In both versions of CrimeWalker taking the strength into account improves the method performance. Comparing to CrimeWalker-DNW and CrimeWalker-CNW, the recall of CrimeWalker-DNS and CrimeWalker-CNS are, respectively, increased by 0.4 and 0.3 %. Although this is not a significant increase but it shows that co-offending links strength has some meaning behind it. Using links strength helps the CrimeWalker method to reach the offenders with higher probability of criminal collaboration in the co-offending network. However, as discussed in Sect. 3.2.2 vast majority of offender pairs only co-offended once and it is not a distinctive feature to improve the method recall significantly.

Using offenders similarity measure enhances the CrimeWalker recall as well. Recall of the CrimeWalker-Disj and CrimeWalker-Conj is increased by 0.5 and 0.6 % comparing to CrimeWalker-DNS and CrimeWalker-CNS. This shows that offenders tend to collaborate with the offenders who are more similar to them. Note that the available features for offenders are very limited which may not provide enough information to compute the offenders similarity precisely.

The other problem is that crime data is usually very noisy and missing data are common. When the police are entering information about offenders into their database, the offenders are sometimes added to the database even when they are unable to provide verifiable information, in which case the police officer will either

Table 5.3 CRIMEWALKER results on DBLP co-authorship network

Method	Recall (%)
CRIMEWALKER-DNW	61
CRIMEWALKER-DNS	64
CRIMEWALKER-CNW	63
CRIMEWALKER-CNS	66
AR	23

fill in the data-field with a default value, such as the address of their own police station for a home location address, or they will leave the field completely empty. In such cases, people who either had default values, or missing data, were considered not to have complete information, and thus were not able to participate in the similarity function, or any of the benefits it may provide. The result was that half of the population either had no data, or default values, thus only approximately 25 % of the links could be assigned similarity values, severely hindering the benefit of the similarity function.

Comparing CRIMEWALKER-Disj and CRIMEWALKER-Conj, we see that CRIMEWALKER-Conj slightly outperforms CRIMEWALKER-Disj. This implies that the method that punishes the recommendations in which one suspect is not highly related to one of the charged criminals achieves higher recommendation quality than the one considering the average among all given charged criminals.

Table 5.3 presents the results on the DBLP dataset. Again, the full model outperforms the other comparison partners. Note that since we have no author attributes, the similarity of nodes cannot be computed and therefore we only have the co-authorship links strength. The CRIMEWALKER performance on the co-authorship network is more than eight times greater than its performance in the co-offending network. A co-authorship network has fundamental differences comparing to a co-offending networks which are the reasons of the significant difference in the CRIMEWALKER performance. For instance, an author always prefers to initiate academic collaborations and publish her results, while an offender always tries to hide her criminal collaborations and activities. This and other reasons make predictive tasks difficult in the crime world comparing to the similar tasks in the other domains.

5.5 Conclusions

Investigating crime can be a challenging and difficult task, especially in cases with many potential suspects and inconsistent witness accounts or inconsistencies between witness accounts and physical evidence.

We present here a novel approach to crime suspect recommendation using a random walk based method for recommending the top-N potential suspects based on partial knowledge of the offenders involved in a crime incident and a known co-offending network. The proposed model, CRIMEWALKER, extends the existing random walk based model, TrustWalker, to address link prediction combined with the ability to perform recommendations based on a set of offenders given as input instead of a single offender. We discuss different ways of how recommendation results for a single offender can be merged into one single top-N result.

Our experimental results, obtained from evaluating CRIMEWALKER on two real-world datasets, BC crime dataset, and DBLP co-authorship network dataset, demonstrate that using the random walk based method on co-offending networks can produce high-quality suspect recommendations. Our experiments also show that using the weighted co-offending network and similarity measure further improves the quality of recommendation.

References

1. R. Agrawal, T. Imieliński, A. Swami, Mining association rules between sets of items in large databases. ACM SIGMOD Record **22**(2), 207–216 (1993)
2. R. Agrawal, R. Srikant, Fast algorithms for mining association rules in large databases, in *Proceedings of the 20th International Conference on Very Large Data Bases*, VLDB'94, pp. 487–499, 1994
3. K. Baumgartner, S. Ferrari, G. Palermo, Constructing Bayesian networks for criminal profiling from limited data. Knowl.-Based Syst. **21**(7), 563–572 (2008)
4. A.A. Block, *East Side-West Side: Organizing Crime in New York City, 1930–1950* (Transaction Publishers, New Brunswick, 1994)
5. R. Boba, *Crime Analysis and Crime Mapping* (SAGE Publications, Thousand Oaks, 2013)
6. S. Ferrari, K. Baumgartner, G. Palermo, R. Bruzzone, M. Strano, Network models of criminal behavior. IEEE Control. Syst. Mag. **28**(4), 65–77 (2008)
7. X. Fu, J. Budzik, K.J. Hammond, Mining navigation history for recommendation, in *Proceedings of the 5th International Conference on Intelligent User Interfaces (IUI'00)*, pp. 106–112, 2000
8. M. Gori, A. Pucci, Itemrank: a random-walk based scoring algorithm for recommender engines, in *Proceedings of the 20th International Joint Conference on Artificial Intelligence (IJCAI'07)*, pp. 2766–2771, 2007
9. P. Hájek, I. Havel, M. Chytil, *The Guha Method of Automatic Hypotheses Determination*, vol. 1 (Springer, Berlin, 1966), pp. 293–308
10. J. Han, M. Kamber, *Data Mining: Concepts and Techniques*. (Morgan Kaufmann, New York, 2006)
11. M. Jamali, M. Ester, TrustWalker: a random walk model for combining trust-based and item-based recommendation, in *Proceedings of the 15th ACM SIGKDD International Conference on Knowledge Discovery and Data Mining (KDD'09)*, pp. 135–142, 2009
12. M. Jamali, M. Ester, Using a trust network to improve top-N recommendation, in *Proceedings of the 3rd ACM Conference on Recommender Systems (RecSys'09)*, pp. 181–188, 2009
13. P. Lazarsfeld, R. Merton, Friendship as a social process: a substantive and methodological analysis. in *Freedom and Control in Modern Society*. ed. by T. Abel, M. Berger, C. Page (Van Nostrand, New York, 1954)

14. W. Lin, S.A. Alvarez, C. Ruiz, Efficient adaptive-support association rule mining for recommender systems. Data Min. Knowl. Disc. **6**(1), 83–105 (2002)
15. D. McAndrew, The structural analysis of criminal networks. in *The Social Psychology of Crime: Groups, Teams, and Networks*, ed. by D. Canter, L. Alison (Dartmouth Publishing, Hanover, 1999)
16. M. McPherson, L. Smith-Lovin, J.M. Cook, Birds of a feather: homophily in social networks. Annu. Rev. Sociol. **27**(1), 415–444 (2001)
17. M.A. Tayebi, M. Jamali, M. Ester, U. Glässer, R. Frank, CrimeWalker: a recommendation model for suspect investigation, in *Proceedings of the 5th ACM Conference on Recommender Systems (RecSys'11)*, pp. 173–180, 2011
18. Z. Xia, Fighting criminals: Adaptive inferring and choosing the next investigative objects in the criminal network. Knowl.-Based Syst. **21**(5), 434–442 (2008)
19. J.J. Xu, H. Chen, Fighting organized crimes: using shortest-path algorithms to identify associations in criminal networks. Int. J. Dec. Support Syst. **38**(3), 473–487 (2004)

Chapter 6
Co-offence Prediction

In this chapter, we propose a framework for co-offence prediction using supervised learning. Even though supervised learning methods for link prediction have been studied widely [11–13, 23], to the best of our knowledge, there is no study on supervised learning for co-offence prediction.

Contrary to other social networks, concealment of activities and identity of actors is a common characteristic of co-offending networks. Still, the network topology is a primary source of information for co-offence prediction. Besides, there are two other major information sources: environmental activity and criminal activity. Offenders who are spatially close are socially close too as this increases the chance of meeting each other and forming new criminal collaborations [21]. Further, common criminal experience (with the same type of offences) also affects co-offending behaviour [25].

The proposed framework builds on criminological theories [2, 5, 15, 18, 20] and, considering the available information on offenders, distinguishes three different criminal cooperation opportunities: socially related, geographically related and experience-related. We study the co-offence prediction problem in each of these prediction spaces separately, achieving two goals. First, the heavy class imbalance between positive (existing links) and negative samples (non-existing links) is the main challenge of the link prediction problem [13]. The restriction of the training and test data to the different prediction spaces reduces the class imbalance ratio significantly, while keeping about half of the positive samples (co-offences). Second, the prediction spaces enhance the understanding of co-offence patterns in different criminal cooperation opportunities.

We define the prediction features in four different categories, *social, geographic, geo-social*, and *similarity*, and evaluate their prediction strength both individually and as a set. Social features indicate social closeness of offenders based on their position in a co-offending network. Geographic features show spatial proximity of offenders based on their residential locations and the location of offences they have committed. Geo-social features combine social and geographic characteristics

M.A. Tayebi, U. Glässer, *Social Network Analysis in Predictive Policing*,
Lecture Notes in Social Networks, DOI 10.1007/978-3-319-41492-8_6

of offenders. Finally, similarity features capture homophily-based characteristics of offenders. Evaluating features individually and also as a set shows that the geo-social features we define outperform other features.

Several studies show that supervised link prediction approaches outperform unsupervised methods [11, 13] that use only topological features [12]. In contrast to unsupervised methods, supervised learning methods can overcome the class imbalance problem [13]. Exploiting the geographic information provided by location-based social network services, some recently proposed link prediction methods consider spatial characteristics of users [19, 24]. In [19], the authors use information about places visited by users, in addition to their social network features, to define prediction spaces which reduce the class imbalance ratio and improve the prediction performance.

Co-offending networks are spatially embedded similar to location-based social networks. However, the environmental effects on the formation of co-offence links and accordingly our approach in defining offenders' spatial closeness are different from those in location-based social networks [6, 19, 24, 26]. The proposed supervised learning framework aims at the public safety and security sector. Although there has been significant research on the reasons for involvement of single offenders in crime [2, 17, 20], there is no comprehensive study on the causes for offender collaboration. We argue that our study on co-offence prediction in different criminal cooperation opportunities opens up a new door to the understanding of co-offence patterns.

The co-offence prediction framework proposed in this chapter aims at advancing the state of the art in crime data mining by making the following contributions: (1) Defining co-offence prediction spaces to reduce the class imbalance; (2) Introducing novel prediction features for co-offence prediction; and (3) Experimentally evaluating the proposed approach on large real-world crime data. Some of our main findings in this research include: (a) Features evaluation has important implications. For instance, preferential attachment is a strong predictor compared to the features extracted from common friends in the network. This implies that the chance of criminal collaboration increases more with the opportunity to commit crimes than with trust or transitivity in the co-offending network. Crime location distance is better predictor than home location distance, meaning that being criminally active in the close districts causes new criminal collaboration; (b) Geo-social features are better co-offence predictors than geographic and social features. This result implies that we need to focus more on combined patterns in environmental and social features to enhance crime reduction and prevention; (c) The experimental results show that, although there is variability in the performance of different classifiers, the probability of predicting a co-offence for similarity-related offenders is higher than for socially and geographically related co-offenders.

In Sect. 6.1 related work is reviewed. Section 6.2 briefly introduces basic concepts. Section 6.3 then explains the proposed co-offence prediction approach, and Sect. 6.4 describes the prediction feature sets. Next, Sect. 6.5 presents our experimental evaluation and results. Finally, Sect. 6.6 concludes this chapter.

6.1 Background

This section discusses published works on crime prediction and link prediction.

6.1.1 Crime Prediction

Two main criminology theories claim that involvement in crime is the result of: (1) an individual's crime propensity; and (2) criminogenic features of the environment to which an individual is exposed. While propensity towards crime has long been studied, in the last few decades criminogenic features of the environment received specific attention. Apart from individuals, spatial aspects increasingly gain momentum, and environmental criminology [2] plays an essential role in crime reduction and prevention tactics. New research areas emerge, like crime mapping [10], geographic profiling [18] and crime forecasting [8, 14], that support growing and imperative applications of this research field for law enforcement and criminal intelligence agencies.

Crime prediction methods in the literature completely ignore the role of co-offending in committing crime. Rather their goal is modeling observed crimes spatially and temporally to predict the time and location of future crimes. For instance in [14], the authors use a point-pattern-based transition density model for crime space-event prediction considering criminal preferences from previous crimes.

Given partial information about a crime incident, in Chap. 5 we propose CRIMEWALKER an unsupervised method for top-N suspect recommendation, which applies a random walk based method on co-offending network. In the suspect investigation problem the goal is finding offenders who most likely collaborated with a set of charged offenders. For this purpose CRIMEWALKER computes the probability of collaboration of pair of offenders given their characteristics, and aggregate these probabilities to recommend the top-N suspects. But in the co-offence prediction problem the goal is computing the probability of every pair of offenders who have not collaborated yet considering the patterns of all co-offences in the training data.

As concluded in the link prediction literature, supervised methods always outperform the unsupervised competitors. To the best of our knowledge, the proposed approach in this chapter is the first supervised learning framework for co-offence prediction.

6.1.2 Link Prediction

Link prediction is an important aspect of social network analysis to better understand the network structure. Link prediction methods can be used to extract missing information, identify hidden interactions, evaluate network evolution mechanisms, and beyond.

Most unsupervised link prediction methods [12] rely solely on the network topology and assign scores to potential links based on structural proximity measures such as node neighborhoods or path information. On the other hand, any classification method can be used for link prediction [11, 13, 23]. In a detailed study [13], Lichtenwalter et al. examine key factors in the link prediction problem, and propose a framework for supervised link prediction.

More recently, location-based social networking services made it possible to study and predict spatial behaviours of social network actors. Wang et al. [24], using trajectory and communication patterns of users, concluded that combining mobility and network features enhances the link prediction results. Scellato et al. [19], based on their study of the link prediction problem in online location-based social networks, conclude that using information about places people visited boosts the link prediction performance.

Although the challenges and concerns of our study are similar to other link prediction studies [11, 13, 19, 24], there are fundamental distinctions as well. Crime data and co-offending network characteristics are inherently different from affiliation networks, such as co-authorship networks or mobile phone networks. Although environmental activity plays an important role in creating new links, these activities are also very different from location-based social networks. Therefore the feature or prediction spaces defined based on criminological theories are different from the ones known in the literature. After all, the co-offence prediction framework proposed in this chapter and supported by experimental evaluation is an innovative application of social network analysis that offers a new perspective for crime reduction and prevention strategies.

6.2 Concepts and Definitions

In this section, we explain basic concepts and definitions.

6.2.1 Notations

For co-offending network $G(V,E)$ and $u \in V$, let Γ_u^n be the subset of offenders in V such that their shortest path distance from u is n. $P_u = (p_u^1, p_u^2, \ldots, p_u^K)$ denotes the frequency of crimes committed by u for each of k crime types in the dataset. The subset of offenders who have committed p crime types that coincide with crime types committed by u is referred to by Θ_u^p.

For u in $G(V,E)$, let $H_u = \{h_u^1, h_u^2, \ldots, h_u^k\}$ denote the known home locations of u, and $C_u = \{c_u^1, c_u^2, \ldots, c_u^l\}$ denote the known crime locations for all offences committed by u. Finally, let $\Delta_u = \{\delta_u^1, \delta_u^2, \ldots, \delta_u^k\}$ denote the time periods (intervals) $\delta_u^i = [t_u^{i_1}, t_u^{i_2}]$ that u lived in each of the k home locations given by H_u. $D(h_u^i, h_v^j)$ is the geodesic distance between h_u^i and h_v^j.

6.2.2 Offenders' Activity Space

The *activity space* of an offender has two main components: *Nodes* and *Paths*. Activity Nodes refer to locations a person frequently visits in the course of daily routine activities. Activity Path represents a common route for a routine trip to a frequently visited location. *Crime Pattern Theory* [3] contends that, rather than venture into new areas to commit crimes, offenders act on criminological opportunities arising in areas they are familiar with, more specifically, areas that are part of their activity space. We discuss more in depth about activity space in Chap. 7.

Inverting research on crime pattern theory [2], geographic profiling [18] seeks to either estimate the location of the residence of an offender or to extrapolate locations where an offender is likely to commit future crime. Centrography is one of the most common search approaches for criminal investigation [18]. In [18], Rossmo shows that some offenders live close to the centroid of their crime locations. The "Circle Hypothesis" by Canter et al. [4] defines a circular area around the location of the first offence in a series of crimes to indicate the general area of an offender's home location.

In this research our goal is using offenders' environmental activities for co-offence prediction, while it does not fit with the bases of works in geographic profiling. Considering the definitions in the literature and the general limitations in the crime data we use a simple and safe definition in this regard. For a given offender u with home locations H_u, $A_u^R = \{a_u^1, a_u^2, \ldots, a_u^k\}$ states the activity space of u, where a_u^i is defined as a circle of radius R with h_u^i at its center. We define the *common activity space* of two offenders as the area in which both are active simultaneously, if their activity spaces overlap at any time. For offenders u and v, $A_{u,v}^R = \{a_{u,v}^{1,2}, a_{u,v}^{1,3}, \ldots, a_{u,v}^{p,k}\}$ is their common activity space, that is

$$A_{u,v}^R = \{a_{u,v}^{i,j} | \, a_u^i \cap a_v^j \neq \varnothing \wedge \delta_u^i \cap \delta_v^j \neq \varnothing\}.$$

Intuitively, $a_{u,v}^{i,j}$ represents the intersection of the activity space of u and v during a time they were both active at a_u^i and a_u^j. The offenders who live in $a_{u,v}^{i,j}$ and the crimes in this area are referred to by the set $\chi_{u,v}^{i,j}$ and the set $\phi_{u,v}^{i,j}$, respectively.

To complement our simple definition of activity space, in Chap. 7 we propose an approach to detect offenders' activity space in a probabilistic way.

6.2.3 Geographic and Network Proximity

Neighborhood greatly influences the formation of communities and social networks. Hence, social problems are often studied in connection with neighborhoods. Many researches suggest that crime is as well strongly linked to geographical characteristics. Criminology researches use spatial analysis of crime to understand the

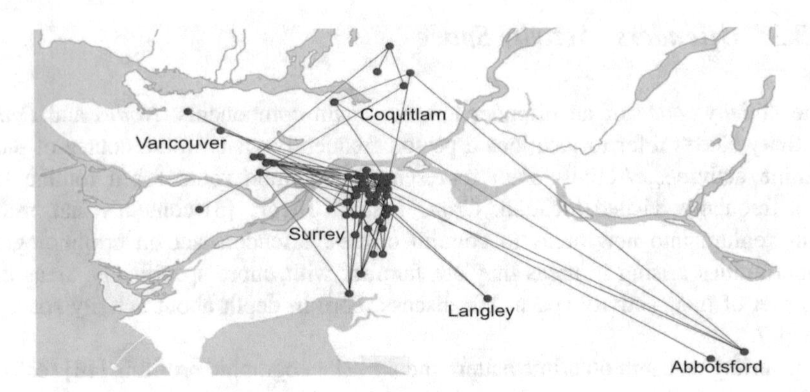

Fig. 6.1 Spatial distance of co-offenders in the GVRD

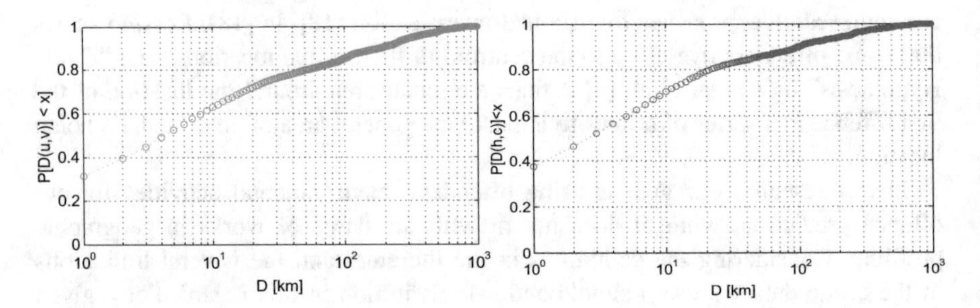

Fig. 6.2 (**a**) Home location distance of co-offenders; (**b**) Home and crime location distance of offenders

distribution of crime incidents and why crime occurs in some places but not others [2, 18]. Figure 6.1 visualizes home locations of offenders for one of the connected components of the BC co-offending network. The fact that clusters can be seen in this visualization implies that many pairs of co-offenders live in the same city.

Generally, it is important to take into account the spatial dimension of co-offending relationships. We have studied this aspect for all observed criminal collaborations in the BC crime dataset. About 39 % of the co-offenders live less than 2 km apart, and about 63 % of them live less than 10 km apart. Figure 6.2a shows the probability distribution of home location distance for co-offenders. The probability distribution of home and crime location distance of offenders is shown in Fig. 6.2b. Forty six percent of the crime incidents happen in less than 2 km distance from the home location, and 70 % of the crimes happen within a distance of less than 10 km. We conclude that a large percentage of the crime incidents are located close to the offenders' residence. While the maximum values for co-offenders home location distance and offenders' home and crime location distance are about 1000 km, the medians of these distances are only 5 km and 3 km, respectively. This confirms that co-offenders tend to be geographically confined.

In [21], it is concluded that socially close offenders are spatially close too, meaning that offenders who are close in the co-offending network have more overlap in their activity spaces too. Criminological theories and the discussed experimental results motivated us to focus more on geographic and geo-social features, besides general social features extracted from network topology for co-offence prediction.

6.2.4 Problem Definition

A co-offending network $G(V,E)$ refers to all co-offences within time period $[t_0,t_n]$ associated with the underlying crime dataset. For any time $t \in [t_0,t_n]$, one can derive $G_t(V_t,E_t)$ as a substructure of G, which is a co-offending network, by restricting to all crime incidents up to time t. For $G_t(V_t,E_t)$ we now define a *potential co-offence* at time $t+1$ as any pair of offenders (u,v) such that $(u,v) \notin E_t$, meaning that offenders u and v have not committed any crime together prior to $t+1$. A potential co-offence (u,v) is in the *positive class*, if $(u,v) \in E_{t+1}$, and it is in the *negative class*, if $(u,v) \notin E_{t+1}$. The *co-offence prediction* task [22] is to predict for each potential co-offence in G_t if it belongs to the positive class or the negative class.

Co-offence prediction can be viewed as a link prediction problem for co-offending networks. This aspect is modeled by means of a binary classification problem that adopts a set of prediction features as described in Sect. 6.4. The major challenge in any link prediction problem is the heavily skewed distribution of negative and positive classes. Contrary to unsupervised methods, supervised methods learn class distributions to increase classification performance [13].

6.3 Supervised Learning for Co-Offence Prediction

In this section, we propose a novel analytic framework for solving the co-offence prediction problem [22].

6.3.1 Criminal Cooperation Opportunities

Because of all obvious concerns about committing crimes, offenders do not select their collaborators accidentally. Considering offenders' choices and selections opportunities, any co-offence can have social, environmental, or experience related cause. In order to quantify how offenders form a criminal cooperation, for each offender u we define three sets of potential co-offenders:

We distinguish three basically different opportunity spaces for criminal cooperation defined as follows:

Socially-Related Social interactions influence the behaviour of individuals. Learning illegal behaviour associated with criminal activities depends on informal networks and peer interactions. Sutherland, in the theory of *differential association* [20], explains individual criminality with a social-psychological process of learning crime through interaction with social groups. According to [20], criminal behaviour is the result of learning an excess of definitions favorable to crime. Offenders use social interactions for sharing information, recruiting young criminals [17] and transferring skills. Although co-offending networks are generally considered short-lived networks, they constitute an important source of criminal cooperation opportunities. In our definition, socially related cooperation opportunities for an offender u, denoted by S_u, arise from offenders v such that their distance in the co-offending network is not greater than N, excluding the direct neighbors of u:

$$S_u = \left\{ (u,v) : v \in \left(\bigcup_{i \leq N} \Gamma_u^i \right) \setminus \Gamma_u^1 \right\}$$

Geographically-Related Activity space and living environment of offenders have major effects on their criminal decisions and are key factors in forming collaboration opportunities [2, 7]. Offender u is considered geographically related to any offender v, if they have a common (overlapping) activity space, but they are not connected to each other directly in the co-offending network.

$$G_u = \left\{ (u,v) : v \in \left(\bigcup_{a_u^i \in A_u^r} \chi_u^i \right) \setminus \Gamma_u^1 \right\}$$

Experience-Related Network studies on offenders have generally observed evidence of homophily such as criminal experience [25]. This is so consistent that group homogeneity is considered as one of the key characteristics of co-offending [25] Having similar criminal experiences increases the chance of forming new criminal collaborations between offenders. This can happen via implicit networks or due to demand for specific criminal expertise. All pairs of offenders who have similar criminal experience but are not directly connected to each other in the co-offending network are categorized into this class:

$$E_u = \left\{ (u,v) : v \in \left(\bigcup_{i \geq P} \Theta_u^i \right) \setminus \Gamma_u^1 \right\}$$

We use the above criminal cooperation spaces to define the spaces which cover the prediction candidates. Prediction space division has two advantages: First, it helps to reduce the class imbalance ratio. Second, one can gain a clearer understanding of the effects of each of these categories on new co-offence formation. For $G_t(V_t, E_t)$ we generate these three spaces for the co-offence prediction task:

SR This space includes all links emerging between a pair of offenders that are socially related: $SR^t = \{\forall u \in V_t : \bigcup_u S_u\}$. In our experiments for constructing the SR space, we consider $N = 2$. In other words, a pair of offenders who are exactly 2-hops apart are added to this space.

GR that includes all links between offenders who are geographically related but not socially related: $GR^t = \{\forall u \in V_t : (\bigcup_u G_u) \setminus S_u\}$. For constructing the activity space of offenders which defines if two offenders are geographically related we use $r = 2\,\text{km}$.

ER that contains all pairs of offenders that are experience-related but not socially related: $ER^t = \{\forall u \in V_t : \bigcup_u E_u \setminus SR_u\}$. For creating the ER space, we apply $P = 2$, meaning that (u, v) are considered experience-related offenders if both committed at least two crimes of the same type.

6.3.2 Reducing Class Imbalance Ratio

For a network $G(V, E)$ the number of links E is often $O(|V|)$, but the number of no-existing links is often $O(|V|^2)$. Accordingly, the prior probability of link formation is very small. Since the goal of supervised learning is achieving high precision, class imbalance in the training dataset leads to overfitting to negative samples which reduces the recall of positive samples. Note that in the co-offence prediction task high recall of the positive class (co-offences) is critical, so that overcoming the class imbalance is essential.

In the experimental setting defined in Sect. 6.5.1, the number of negative samples is 850 M while the number of positive samples is only 11 k. Therefore, the *class imbalance ratio*, the ratio of negative samples to positive samples, is about 77 K. Using the prediction spaces to reduce the training and test dataset can effectively reduce the class imbalance. In [19] an efficient prediction space division schema for location-based social network is proposed, but it does not apply to co-offence prediction where environment activity of offenders is different than location-based social network users. Furthermore we consider the homophily effects, in terms of criminal activity similarity, on forming new link which is not considered in the approach proposed in [19]. Although reducing class imbalance is an important objective, the other main concern is keeping as many positive samples as possible.

Restricting the dataset to samples with short graph distance is the most effective solution for the imbalance problem. In the SR space increasing N affects the number of negative samples set superlinearly, because we expect to see more co-offences between offenders which are close in the co-offending network. Figure 6.3a demonstrates the probability of a positive sample in the SR space for different values of N of the network distance. The probability does not grow linearly as N increases,

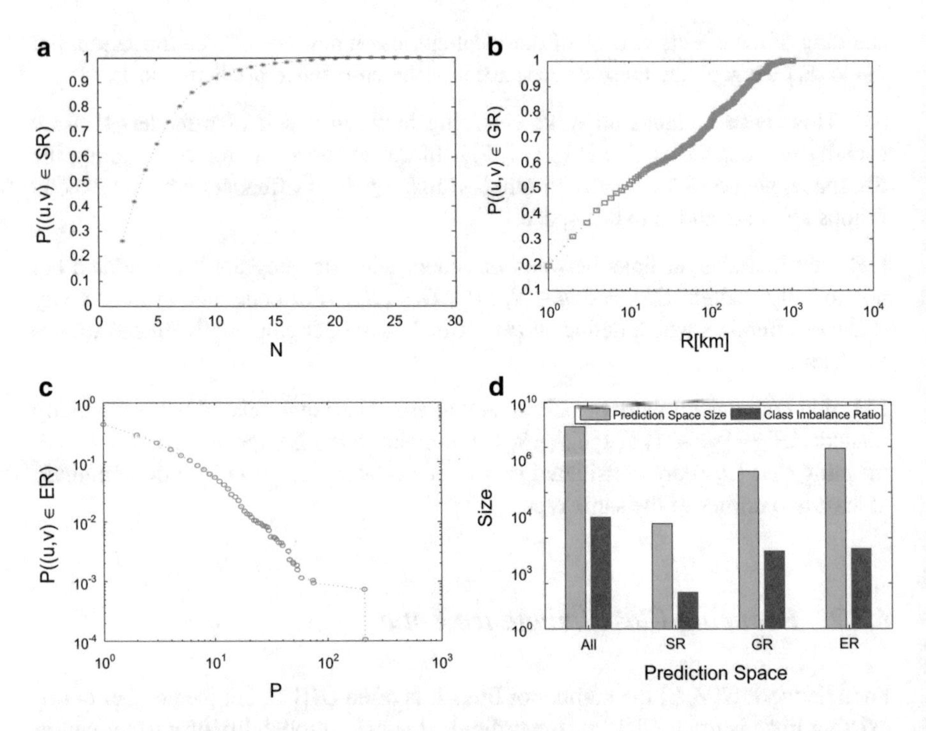

Fig. 6.3 (**a**) Cumulative probability distribution of a positive sample in SR space in respect of N; (**b**) Cumulative probability distribution of a positive sample in GR space in respect of R; (**c**) Cumulative probability distribution of a positive sample in ER space in respect of P; (**d**) Prediction spaces size and their class imbalance ratio

meaning that similar to other type of social networks [13, 19] majority of positive samples are at a closer network distance. For the maximum value $N = 27$ there are 204 M negative samples. But changing this to $N = 2$ decreases the number of negative samples to 47 K, while keeping 25 % of the positive samples.

Applying different constraints related to the other prediction spaces also helps to reduce the class imbalance ratio. Figure 6.3c shows the probability of a positive sample in the GR space for different values R of the radius of the activity space. With $R = 2$, $R = 10$, and $R = 100$ km the GR space covers cumulative 31, 51, and 75 % of the positive samples. Figure 6.3b shows the probability of a positive sample in the ER space for different values of P. With $P = 2$ of the same crime type, $P = 5$ and $P = 10$ km the ER space includes 29, 13, and 5 % of the positive samples. Applying $R = 2$ km and $P = 2$ reduces the 850 M negative samples in the dataset to 8 and 82 M, respectively, for GR and ER spaces. With these constraints we can keep 31 and 29 % of the positive samples in the GR and ER spaces.

As demonstrated in Fig. 6.3d, the original imbalance ratio of 77 K for the whole prediction space reduces to 40, 2700, and 3400, respectively, for SR, GR, and ER spaces. While in each of three spaces roughly we can keep approximately the same percentage of positive samples (26, 31, and 29 % for the SR, GR, and ER), we see that the class imbalance reduction is better in the SR space. In total we are able to maintain half of positive samples for training, meaning that the likelihood of predicting a co-offence successfully increases significantly.

6.4 Prediction Features

Prediction features are divided into four categories: social, geographic, geo-social, and similarity features. This section describes how these features are extracted [22]. Features are defined formally in Table 6.1.

6.4.1 Social Features

The social features set includes all features that are derived using only the topology of the co-offending network and the position of offenders in the network. Preferential is defined as the product of the node degrees of two offenders, meaning that the more connected an offender is, the more likely he forms new criminal links. Common denotes the number of neighbors two offenders have in common. Overlap divides the number of common neighbors by the overall number of neighbors of two offenders as a normalized indication of common neighbors effect. Adamic [1] assigns higher weight to the common neighbors with smaller node degree.

6.4.2 Geographic Features

With increasing overlap of the activity space of offenders the chance of forming new criminal collaboration increases. Home locations distance, HDN, is the average distance between the current and past home locations of two offenders. HDT weights home location distance by the time two offenders lived in the corresponding locations. Committing crimes in the same neighborhood may cause a criminal tie between offenders. CDN is the average crime location distance of two offenders.

6.4.3 Geo-Social Features

Geo-social features combine the social and geographical characteristics of offenders. Common activity space of offenders directly affects their behaviours. Being active in an area with many offenders living there can increase the chance that offenders meet each other directly or indirectly and engage in co-offending. OCT denotes the number of offenders living in a common activity space of two given

Table 6.1 Prediction features

Social features	
Preferential	$\lvert\Gamma_u^1\rvert \times \lvert\Gamma_v^1\rvert$
Common	$\lvert\Gamma_u^1\rvert \cap \lvert\Gamma_v^1\rvert$
Overlap	$\frac{\lvert\Gamma_u^1\rvert\cap\Gamma_v^1}{\lvert\Gamma_u 1\rvert\cup\lvert\Gamma_v^1\rvert}$
Adamic	$\sum_{z\in\Gamma_u^1\cap\Gamma_v^1} \frac{1}{log(\Gamma_z^1)}$
Geographic features	
HDN	$\frac{\sum_{i=1}^{i=m}\sum_{j=1}^{j=n} e^{\frac{-D(h_u^i,h_v^j)}{\lambda}}}{\lvert H_u\rvert\times\lvert H_v\rvert}$
HDT	$\frac{\sum_{i=1}^{i=m}\sum_{j=1}^{j=n} e^{\frac{-D(h_u^i,h_v^j)}{\lambda}} \times\lvert(\delta_u^i\cap\delta_v^j)\rvert}{\lvert H_u\rvert\times\lvert H_v\rvert}$
CDN	$\frac{\sum_{i=1}^{i=m}\sum_{j=1}^{j=n} e^{\frac{-D(c_u^i,c_v^j)}{\lambda}}}{\lvert C(u)\rvert\times\lvert C(v)\rvert}$
Geo-social features	
OCT	$\sum_{i=1}^{i=p}\sum_{j=1}^{i=k}\lvert\chi_{u,v}^{i,j}\rvert$
& OCTT	$\sum_{i=1}^{i=p}\sum_{j=1}^{i=k}\lvert\chi_{u,v}^{i,j}\rvert\times\lvert\delta_u^i\cap\delta_v^j\rvert$
OCN	$\sum_{i=1}^{i=p}\sum_{j=1}^{i=k}\lvert\chi_{u,v}^{i,j}\rvert : [t_0,t]$
CCT	$\sum_{i=1}^{i=p}\sum_{j=1}^{i=k}\lvert\phi_{u,v}^{i,j}\rvert$
CCTT	$\sum_{i=1}^{i=p}\sum_{j=1}^{i=k}\lvert\phi_{u,v}^{i,j}\rvert\times\lvert\delta_u^i\cap\delta_v^j\rvert$
CCN	$\sum_{i=1}^{i=p}\sum_{j=1}^{i=k}\lvert\phi_{u,v}^{i,j}\rvert : [t_0,t]$
Similarity features	
Age	$\lvert Age(u)-Age(v)\rvert$
Gender	$\begin{cases}1, & \text{if Gender}(u)=\text{Gender}(v)\\ 0, & \text{if Gender}(u)\neq\text{Gender}(v)\end{cases}$
Ethnic	$\begin{cases}1, & \text{if Ethnic}(u)=\text{Ethnic}(v)\\ 0, & \text{if Ethnic}(u)\neq\text{Ethnic}(v)\end{cases}$
CrimSim	$\frac{\sum_{i=1}^{K} P_u^i P_v^i}{\sqrt{\sum_i^K (P_u^i)^2}\times\sqrt{\sum_{i=1}^K (P_v^{i=1})^2}}$

offenders. In the definition of common activity space time is implicitly restricted to $\delta_u^i \cap \delta_v^j$. OCTT boosts OCT if two offenders are active for a longer period of time. Without any restriction, OCN denoted the number of offenders who live in a common activity space anytime in the time interval $[t_0, t]$.

Characteristics related to the quantity of crimes in the common activity space provide another set of geo-social features. A crime hotspot located within the common activity space of offenders may increase the chance of collaboration among these offenders. CCT states the number of crimes in the common activity space of two offenders, where, similar to OCT, the time is restricted to the period over which both offenders are active simultaneously. CCTT weights the time interval of two offenders being active. And CCN refers to the total number of crime incidents in the common activity space.

6.4.4 Similarity Features

The homophily principle states that individuals tend to associate themselves with others in a social network who are similar [16]. This is also a very well-studied topic for co-offending [25]. Age difference, ethnic group similarity and gender similarity, represented by Age, Ethnic and Gender, are three features in this category. Finally, CrimSim also expresses similarity of criminal experience based on committed crime types.

6.5 Experiments and Results

In this section, we present the experimental evaluation.

6.5.1 Experimental Design

The special nature of co-offending networks makes the co-offence prediction a difficult task. Offenders not only do hide their illegal activities but their accomplice relationships are usually short-lived [17].

For our experiments [22], we divide the dataset into two disjoint sets of incidents, one for the first 50 months and the second one for the last 10 months. We use this setting to include large enough portion of data in the training set to learn criminal collaboration patterns, and the most possible positive samples in the test set. Excluding noisy data, such as traffic related offenses, the number of incidents for the aforementioned time periods are 1.8 M and 800 K, respectively. Considering only offences with more than one offender reduces these numbers to 67 and 17 K. We extract the co-offending networks corresponding to each of these time periods.

Next, criminal cooperation opportunities are identified and each pair of potential co-offences is assigned to one of the three prediction spaces: SR, IR, or ER. Eventually, the prediction features of each pair of potential co-offenders are extracted.

We use the open-source machine-learning library Weka [9] for different classifier building. The performance of classifiers and single features prediction are assessed based on the Receiver Operating Characteristics (ROC), which shows the trade-off between the true positive rate over the false positive rate. The area under the ROC curve (AUC) is an appropriate measure for comparing two ROC curves. AUC measures the probability that a classifier ranks a randomly chosen positive sample higher than a negative sample.

While there are various definitions of offender activity space in the literature, based on the available information on offenders in the dataset, we use a simple definition: a circle of radius R with the home location of an offender in the center of the circle. Common activity space of two offenders is defined as the intersection of their activity space and is used for defining geo-social features. As there is no consensus in the criminology literature about the appropriate value of R we tried different values of R, ranging from 1–10 km. Interestingly, for all features $R = 2$ km maximize the prediction performance, which we therefore choose in our experiments.

6.5.2 Single Features Significance

We compare the significance of different features described in Sect. 6.4 in each of the prediction spaces [22]. This is important to diagnose which features in each prediction space play a more important role in co-offence prediction. For this purpose, we compute the feature values for all negative and all positive potential co-offenders. Then, using a range of decision thresholds and computing the false/true positives ratios, we generate the ROC curves for each single feature as presented in Fig. 6.4c. Note that the social features can be extracted only for SR space.

As shown in Fig. 6.4a, in the SR space, the Preferential feature is the best predictor with AUC value 0.82 and is superior to other social features. The performance of Adamic and Jaccard is worse than a random predictor. This shows that being a frequent offender and having broader criminal relationships, rather than common relationships, increases the chance of engaging in new criminal cooperations. Following Preferential are geo-social and geographic features CDN, OCN, CCN, and HDN, with AUC values between 0.74 and 0.79. The similarity performance of features is lower than the mentioned set.

In the GR space, performance of most of the features is weaker than their performance in the SR space. This is not unexpected for geographic or geo-social features, where in the negative potential class we gather all pairs that are geographically close and the predictor cannot be as successful as in the SR space. Nevertheless, the AUC values of these features fall within the range 0.60–0.70.

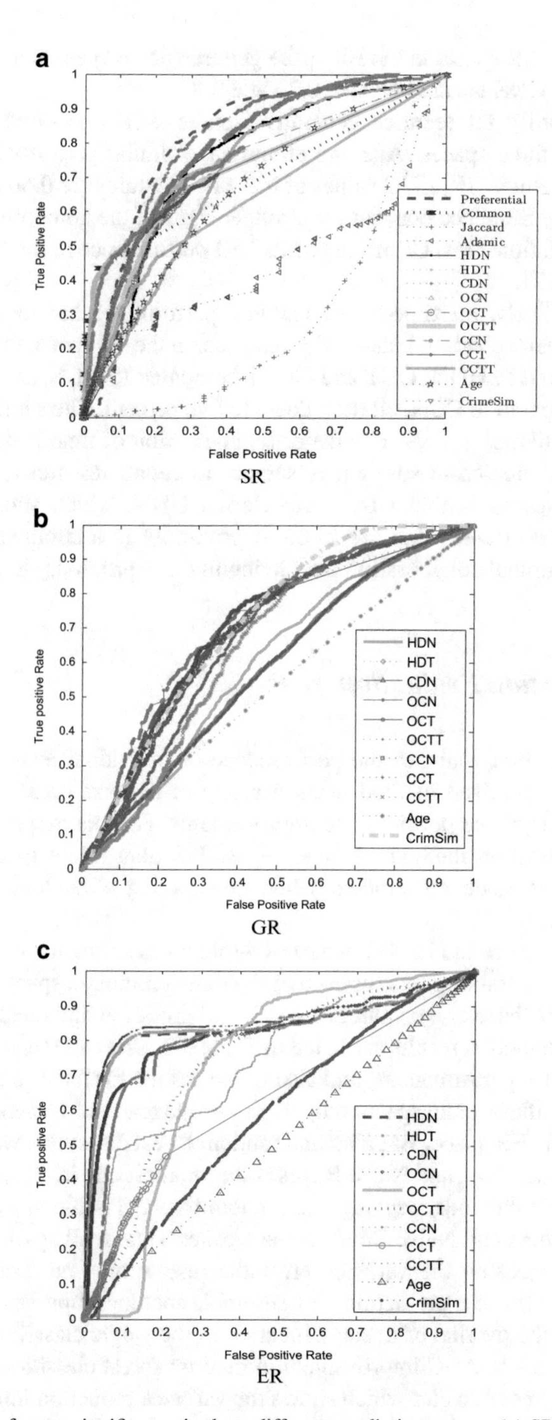

Fig. 6.4 Single feature significance in three different prediction spaces. (**a**) SR, (**b**) GR, (**c**) ER

Compared to the SR space, in the ER space geographic and geo-social feature work better. Their AUC values are between 0.72 and 0.86.

In the similarity set features generally Gender works as good as a random predictor in all three spaces. Age and ethnic have similar performance in the ER and GR spaces. But in SR, AUC values of these two features are 0.66 and 0.60. This shows that in the SR space one can see stronger signs of the homophily principle in forming new relationships. CrimSim has its best performance in the GR space with AUC value of 0.71.

Counterintuitively, all time-based features perform weaker than the original version of the corresponding feature. For instance, in the SR space, the performance of HDT compared to HDN, OCT and OCTT compared to OCN, and, finally, CCT and CCTT compared to CCN, all drop down to five percent. The same trend applies to the other prediction spaces. Parameter regularization of time-based features for promoting their prediction strength is subject to future research. An interesting pattern in all spaces is that CDN outperforms HDN, which shows that being criminally active in areas that are in close proximity is a stronger indicator for forming new criminal collaboration than living in close proximity to each other.

6.5.3 Prediction Evaluation

The previous section analyzes the performance of individual features. Now, how good works a supervised method using a group of extracted features? Naturally, one would expect to see performance improvements. For this purpose, we use four different classification methods: Naïve Bayes, J48 (equivalent to C4.5), random forests (ten trees, each constructed while considering 4 random features), and bagging (ten bags).

Similar to the work in [11, 19], we run 10-fold cross validation over 10 different randomly sampled training sets for each of the three prediction spaces, GR, SR, and ER. We consider the average values of AUC, and precision and recall over positive samples. The prediction results are listed in Table 6.2. One can see variability in the different classifier performances, and also in the results for the different prediction spaces. All classifiers for all spaces outperform single features. Generally, prediction works best in the ER space. Bagging and random forest classifiers work better than the other two classifiers, and Naïve Bayes is the weakest one in all spaces.

Two ensemble methods, bagging and random forest classifiers, work better than the other classifiers and Naïve Bayes is the weakest one in all spaces. In ensemble learning for decreasing the variance error the results of a number of classifiers are combined to make a prediction. An ensemble method improves the prediction performance if the overlap of misclassification of the single classifiers is small. On average we see 5.1 % AUC improvement in random forest classifiers comparing to single decision tree classifier which shows the variance reduction influence.

Since all features are available in the SR space and the bagging classifier works better than other methods, we continue the following experiments with the SR

Table 6.2 Recall, precision and AUC for different classification algorithms on the three different prediction spaces, SR, GR, and IR, using 10-fold cross validation over 10 different random training sets

Algorithm	Space	Precision	Recall	AUC
J48	SR	0.888	0.807	0.907
	GR	0.869	0.834	0.901
	ER	0.935	0.81	0.898
Naïve Bayes	SR	0.836	0.514	0.825
	GR	0.825	0.441	0.817
	ER	0.945	0.706	0.895
Random Forest	SR	0.898	0.843	0.944
	GR	0.864	0.883	0.944
	ER	0.941	0.944	0.982
Bagging	SR	0.908	0.84	0.951
	GR	0.863	0.854	0.952
	ER	0.946	0.942	0.986

Table 6.3 Prediction strength of different feature sets for SR space using the bagging classifier with 10-fold cross validation over 10 different random training sets

Features set	Precision	Recall	AUC
Social	0.903	0.792	0.919
Geographic	0.721	0.786	0.811
Geo-social	0.863	0.853	0.942
Similarity	0.849	0.851	0.928
All features	0.908	0.84	0.951

space and bagging classifier. To study prediction strength of each of the feature sets, social, geographic, geo-social and similarity, each time we keep one of the feature sets and evaluate it using the bagging classifier with the same setting. As shown in Table 6.3, the geo-social features outperform the other three sets, and the geographic feature set has the worst performance. Comparing the prediction performance using all features to different subsets shows integrating all features from social, environmental, and personal can contribute to co-offence prediction.

Most existing unsupervised link prediction methods assign scores to the potential links and rank them. The top-N links are categorized as new links, and the remaining as missing links. The evaluation of unsupervised methods is same as single features evaluation in Sect. 6.5.2. Preferential, Jaccard, and Adamic are common unsupervised link prediction methods as addressed in [12]. While Preferential has the highest the AUC value among individual features we see more than 13 % improvement in the bagging classifier result, which means supervised classifiers outperform the unsupervised predictors significantly.

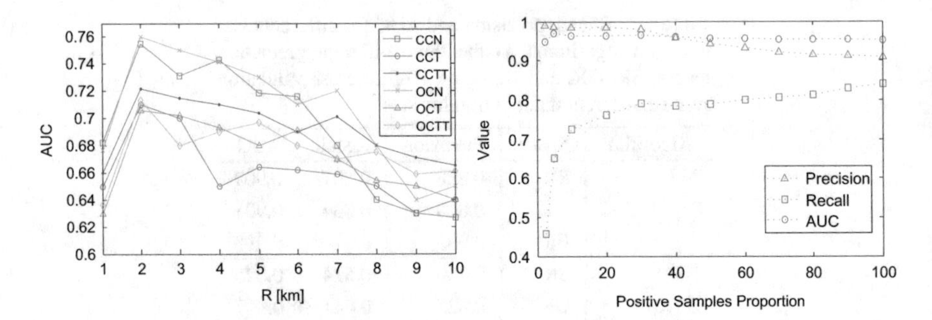

Fig. 6.5 (**a**) Single features significance using different activity space radius(**b**) Performance changes of the bagging classifier for different imbalance ratios. The x-axis states the ratio of the positive class size over the negative class size

For a better insight into class imbalance issues, we vary the size of negative samples for the SR space using same set of positive samples, where the positive samples proportion ranges from 2 % to 100 %. As illustrated in Fig. 6.5b, the AUC value is relatively stable. Unsurprisingly, with negative sample size and accordingly imbalanced ratio growth, precision increases and recall decreases. As expected, this is because of negative class overfitting, meaning that the trained classifier is more likely to predict non-existing links.

6.5.4 *Criminological Implications*

Crime forecasting models are divided into short-term and long-term categories in terms of predicted time periods. Short-term models are beneficial for tactical decision making, whereas long-term models are helpful for planning and policy development. Co-offence prediction is an important aspect for short-term models. Studying new link formation patterns can help designing effective long-term crime reduction and prevention strategies.

With the settings defined in Sect. 6.5.1, $E_{t+1} \cong 120\,\text{K}$, while the portion of positive samples is about 9 %. And in the reality the prediction is possible only in this portion where we had information about both the co-offenders previously. The proposed framework was able to predict 45 % of all positive samples.

Figure 6.6a shows the cumulative degree distribution $P(k)$ in the co-offending network G_{50} for three sets of offenders: V_{50}, $\{u \mid \exists v \in V_t : (u,v) \in \text{SR} \cap E_{t+1}\}$ and $\{u \mid \exists (u,v) \in \text{SR} \wedge (u,v) \notin E_{t+1}\}$. One can see that for the same node degree k, compared to the two other sets, the set of offenders who contribute to forming co-offences generally has a greater value of $P(k)$. Section 6.5.2 shows that Preferential for SR space is a strong predictor. This implies the relevance of co-offending networks for making new criminal cooperation.

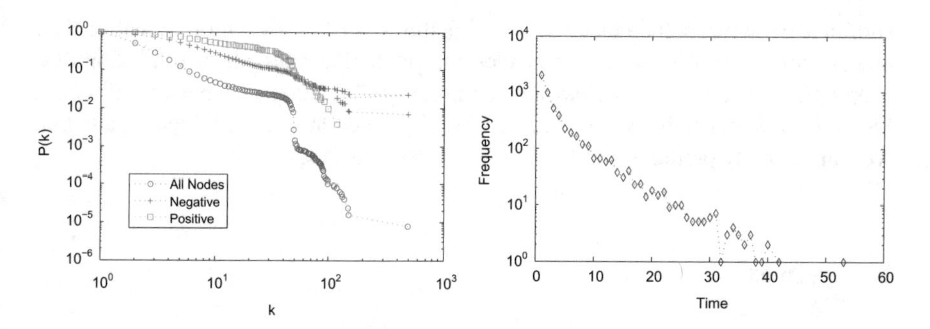

Fig. 6.6 (**a**) Cumulative degree distribution of offenders (**b**) Frequency distribution of time difference between two offenders most recent crime and their first co-offence

Assume that $\tau_{u,v}$ refers to the first time that a positive co-offence (u,v) occurs in time interval $[51,60]$, and τ is the time of the most recent offence of u or v in $[1,50]$. Figure 6.6b plots the distribution of $\tau_{u,v} - \tau$ for all positive potential co-offences. We see a power law distribution, meaning that there are many offences with a small value of $\tau_{u,v} - \tau$ and few with a larger value of $\tau_{u,v} - \tau$. This result shows that offenders who were recently active have a higher potential to form new criminal cooperation.

We see some variance in the prediction strength of different prediction features, with weak and strong predictors in each of the spaces. For all spaces we see high prediction results, which shows that there are strong patterns in co-offending. This supports our idea of dividing criminal cooperation opportunities, this way, finding stronger co-offence patterns that improve predictions. However, co-offending is a product of the intricate interplay between many factors, and drawing any firm conclusions requires more substantial and in-depth research beyond the scope of the work presented in this book.

6.6 Conclusions

This chapter proposes a supervised learning framework for co-offence prediction. We define the co-offence prediction problem as a link prediction problem in co-offending networks. Comprehensively considering criminological theories about social, environmental, and homophily roots of offending, we assign each pair of offenders to socially related, geographically related or experience-related criminal cooperation opportunities, and using these sets we create three prediction spaces. While we are able to retain half of the co-offences, the prediction space division allows us to significantly reduce class imbalance which is a major challenge in link prediction. We do not only use homophily, social and spatial characteristics of pairs of offenders to define prediction features, but we also define novel geo-social features combining social and spatial characteristics. Evaluating single features significance, we conclude that geo-social features generally are more effective than others, while we also observe considerable prediction strength in other features. Employing all features, our classifiers employing all features achieve a recall of 84, 88, and 94 % respectively, for three different prediction spaces, which means that we can correctly predict roughly 90 % of the co-offences.

References

1. L.A. Adamic, E. Adar, Friends and neighbors on the web. Soc. Netw. **25**(3), 211–230 (2003)
2. P.J. Brantingham, P.L. Brantingham, *Environmental Criminology*. (Sage Publications, Beverly Hills, 1981)
3. P.L. Brantingham, P.J. Brantingham, Nodes, paths and edges: considerations on the complexity of crime and the physical environment. J. Environ. Psychol. **13**(1), 3–28 (1993)
4. D.V. Canter, A. Gregory, Identifying the residential location of rapists. J. Forensic Sci. Soc. **34**(3), 169–175 (1994)
5. M. Carlo, *Inside Criminal Networks* (Springer, New York, 2009)
6. E. Cho, S.A. Myers, J. Leskovec, Friendship and mobility: user movement in location-based social networks, in *Proceedings of the 17th ACM SIGKDD International Conference on Knowledge Discovery and Data Mining (KDD'11)* (2011), pp. 1082–1090
7. M. Felson, The process of co-offending, in *Theory and Practice in Situational Crime Prevention*, ed. by M. Smith, D. Cornish (Criminal Justice Press, Monsey, 2003)
8. W. Gorr, R. Harries, Introduction to crime forecasting. Int. J. Forecast. **19**(4), 551–555 (2003)
9. M. Hall, E. Frank, G. Holmes, B. Pfahringer, P. Reutemann, I.H. Witten, The weka data mining software: an update. ACM SIGKDD Explor. Newsl. **11**(1), 10–18 (2009)
10. K. Harries, *Mapping Crime Principle and Practice* (U.S. Department of Justice, Office of Justice Programs, National Institute of Justice, Washington, DC, 1999)
11. M.A. Hasan, V. Chaoji, S. Salem, M. Zaki, Link prediction using supervised learning, in *Proceedings of SIAM International Conference on Data Mining (SDM '06)* (2006)
12. D. Liben-Nowell, J. Kleinberg, The link prediction problem for social networks, in *Proceedings of the 12st ACM International Conference on Information and Knowledge Management (CIKM'03)* (2003), pp. 556–559
13. R.N. Lichtenwalter, J.T. Lussier, N.V. Chawla, New perspectives and methods in link prediction, in *Proceedings of the 16th ACM SIGKDD International Conference on Knowledge Discovery and Data Mining (KDD'10)* (2010), pp. 1100–1108
14. H. Liu, D.E. Brown, Criminal incident prediction using a point-pattern-based density model. Int. J. Forecast. **19**(4), 603–622 (2003)
15. J. McGloin, C.J. Sullivan, A.R. Piquero, S. Bacon, Investigating the stability of co-offending and co-offenders among a sample of youthful offenders. Criminology **46**(1), 155–188 (2008)
16. M. McPherson, L. Smith-Lovin, J.M. Cook, Birds of a feather: homophily in social networks. Annu. Rev. Soc. **27**(1), 415–444 (2001)
17. A.J. Reiss Jr., Co-offending and criminal careers. Crime Justice **10**, 117–170 (1988)
18. D.K. Rossmo, *Geographic Profiling* (CRC Press, Boca Raton, 2000)
19. S. Scellato, A. Noulas, C. Mascolo, Exploiting place features in link prediction on location-based social networks, in *Proceedings of the 17th ACM SIGKDD International Conference on Knowledge Discovery and Data Mining (KDD'11)* (2011), pp. 1032–1040
20. E.H. Sutherland, *Principles of Criminology* (J. B. Lippincott & Co., Chicago, 1947)
21. M.A. Tayebi, R. Frank, U. Glässer, Understanding the link between social and spatial distance in the crime world, in *Proceedings of the 20nd ACM SIGSPATIAL International Conference on Advances in Geographic Information Systems (ACM SIGSPATIAL GIS'12)* (2012), pp. 550–553

22. M.A. Tayebi, M. Ester, U. Glässer, P.L. Brantingham, Spatially embedded co-offence prediction using supervised learning, in *Proceedings of the 20th ACM SIGKDD International Conference on Knowledge Discovery and Data Mining (KDD'14)* (2014), pp. 1789–1798
23. C. Wang, V. Satuluri, S. Parthasarathy, Local probabilistic models for link prediction, in *Proceedings of the 16th ACM SIGKDD International Conference on Knowledge Discovery and Data Mining (ICDM'07)* (2007), pp. 243–252
24. D. Wang, D. Pedreschi, C. Song, F. Giannotti, A. Barabasi, Human mobility, social ties, and link prediction, in *Proceedings of the 17th ACM SIGKDD International Conference on Knowledge Discovery and Data Mining (KDD'11)* (2011), pp. 1100–1108
25. F.M. Weerman, Co-offending as social exchange: explaining characteristics of co-offending. Br. J. Criminol. **43**(2), 398–416 (2003)
26. C. Zhang, L. Shou, K. Chen, G. Chen, Y. Bei, Evaluating geo-social influence in location-based social networks, in *Proceedings of the 21st ACM International Conference on Information and Knowledge Management (CIKM'12)* (2012), pp. 1442–1451

Chapter 7
Personalized Crime Location Prediction

Urban population is expected to grow globally from 2.86 bn in the year 2000 to 4.98 bn by 2030 [48]. While we are within this period of large urban growth, policymakers and law enforcement agencies are facing an enormous challenge deploying scarce policing resources more efficiently and effectively. This effectiveness requires a combination of apprehension of criminals, disruption of criminal networks, and the deterrence of crime through crime prevention and reduction strategies.

The spatial analysis of crime is re-emerging in importance [8, 10, 34, 35, 37, 39, 50]. Studies find that crime does not occur uniformly or randomly across the urban landscapes [4, 8, 27, 37, 50]. Crime *hotspots*, areas with high crime intensity, generate a larger percentage of criminal events [37]. Understanding why hotspots emerge in some places and not in others is a challenging question [5, 6, 8, 37]. Hotspot analysis enables law enforcement to better prioritize their use of resources for crime reduction and prevention. But the concentration of interest on hotspots pulls attention away from better understanding areas with more moderate or low concentration of criminal events. These areas can be referred to as *coldspots*. Better understanding coldspots is of value because these areas account for approximately half of all urban crimes [50].

From the criminological perspective, the best known study of hotspots and coldspots is a 16-year longitudinal study of crime in Seattle Washington in the USA. It finds that roughly half of yearly crime incidents occur within only five to six percent of the city's road segments [50]. Coldspots cover a much wider area than hotspots. Targeted policing, or what is called hotspot policing, is not feasible for crime reduction or prevention in these non-hotspot areas. Better understanding the spatial distribution of crime incidents in coldspots is essential for the development of intervention strategies. In hotspot analysis the focus is on modeling the emergence, evolution and stability of the hotspots. Such analysis is often based on analysis of

M.A. Tayebi, U. Glässer, *Social Network Analysis in Predictive Policing*, Lecture Notes in Social Networks, DOI 10.1007/978-3-319-41492-8_7 aggregate crime patterns. Coldspot analysis, as is explored in this chapter, requires modeling of individual offenders' spatial behaviour. What is needed are models that are flexible and can be personalized to individual offenders.

Existing models of crime distribution mostly focus on models for predicting crime locations for time intervals [15]. These studies rely heavily on modeling hotspot emergence, bifurcation, and diffusion [29, 38] to identify clusters of incidents in crime intensive areas. These models frequently use concepts of crime attractiveness that pull people towards locations [11, 17, 38, 39, 41]. There is some research that explores decision models for offenders following a McFadden decision theory [4] and the tendency of offenders to commit offences near prior offences [26]. But the models tend to use one decision making process for all potential offenders.

The model presented in this chapter focuses on individualized offending and decision making with the decision rules being different for the occasional offender and the frequent repeat offender and repeat co-offender. The model is derived from *Crime Pattern Theory* [10] based on the assumption that offenders, rather than venture into unknown territory, frequently commit crimes in places they are most familiar with as part of their activity space [10]. We used a simple definition of activity space in Sect. 6.2.2, but our proposed method in this chapter generates a probabilistic activity space for every offender.

Activity space is shaped by major activity nodes such as home, work/school and entertainment or shopping locations, that is, the frequently visited places determined by a person's daily routine activities. Surrounding an activity space, an individual develops visually, and through local information, an awareness of the surrounds. An activity space is a subset of an awareness space. Activity spaces and awareness spaces change over time with movement to new home locations, new employment, the development of new shopping and entertainment areas, and the development of new mass transit and roads. But, fortunately, crime is relatively rare, and acceptable targets of crime or victims are likely to be found easily within an awareness space. Outside an activity or awareness space, an offender will have to consciously hunt for criminological opportunities and likely face higher uncertain or unforeseeable risks. *Crime occurrence space* is more likely a part of an activity space that intersects with the location of suitable targets preferred by an offender.

The focus of this model is on crimes that are linked to individual offenders in non-hotspot urban areas. It uses random walk to model how offenders encounter criminal opportunities at a local level near anchor locations in an activity space [40]. In [40], the authors propose a random walk based model for capturing the dynamics of hotspot formation (see [40] for a Levy Flight model). We present here an extended random walk model, CRIMETRACER, for generating the activity space associated with offenders living in an urban area. In CRIMETRACER, the random walk process is personalized to uncover the spatial behaviour of all individual offenders.

For the urban layout we assume a small-scale road network on which an offender moves about in an urban area. By doing so, we gradually compute an approximation of the offender's activity space by reflecting the probability of visiting (and possibly committing crime) for each road segment of the urban area. This result is then used for predicting crime locations for individual offenders, something not addressed in crime spatial analysis to the best of our knowledge. Based on our experimental evaluation, personalization is successful for detecting crime locations in coldspots. The extended random walk model outperforms the random walk model and the other evaluated methods in terms of the recall and precision metrics.

Section 7.1 discusses the related work. Section 7.2 introduces the CRIME-TRACER model. Section 7.3 presents the experimental evaluation. Finally, Sect. 7.4 concludes this chapter.

7.1 Background

The environment that we live in influences our actions and movements in different ways. Highways, streets, and road networks guide us to our destinations. Our destinations including home, workplace, recreation center, and business establishments determine where we are going. Consequently, where we are at a point of time is dictated by our destinations. On the other hand, the transportation routes leading us to our destinations are limited and predictable. Each individual has normal, routine pathways or commuting/mobility routes which are unique. Yet in the aggregate in time and space we often have rush hours and congestion at intersections or mass transit stops that handle large numbers of people.

Offenders are most often "non-offenders" and develop routine mobility patterns and routine alternate routes. In many ways they follow the same process in their mobility as non-offenders in the urban environment. Their spatial awareness is formed based on their destinations and transportation routes, and potential targets located in frequently visited places may attract them.

For predicting crime locations of offenders we need to understand offender mobility patterns. In this section, we describe criminological theories and research that addresses mobility.

7.1.1 Spatial Pattern of Crime

Place is important in bringing offenders and crime targets together. Crime is not distributed uniformly or randomly across a city. Instead, crime is associated with interaction between offender activity space and places with suitable targets. Spatial characteristics play an important role in criminal activity. Environmental criminology concludes that "patterning of crime, and even the volume of crime, depends on motivation and opportunity, and mobility and perception" [9]. Criminological studies often focus on these aspects of crime spatial pattern: Crime Hotspot Analysis for areas with high crime rates and Journey to Crime Analysis focus on the impact of offender mobility on patterns of crime. Journey to crime studies generally consider two factors: direction and distance, and conclude that offender trips are generally, but not always, short and directed to specific locations [7, 47].

Environmental criminology, the field of studying crime in relation to particular places, includes three important theories: routine activity theory, rational choice theory, and Crime Pattern Theory. Routine activity theory focuses on offenders within the larger society and rational choice theory focuses on offender decision making processes. Crime Pattern Theory adds a new dimension: interaction and movement of offender and victim through place [16].

Crime Pattern Theory, with the emphasis on spatial aspects, has become a central theory of crime spatial pattern analysis. Our CRIMETRACER model presented in Sect. 7.2 is constructed based on Crime Pattern Theory.

7.1.2 Crime Pattern Theory

Offenders, just like other people, have their own activity spaces they build up from their regular non-offending portions of their lives, as well as through their offending activities. Crime Pattern Theory [10] says that, rather than venture into new areas to commit crimes, offenders will act on criminological opportunities that are within the areas they are familiar with, more specifically those that are within their activity space. This is intuitively correct. If the offender is outside of their activity space, they will be in unfamiliar territory and will have to hunt for a criminological opportunity.

In one sense, looking for targets outside of a known area may mean passing suitable targets in a known area with the hope that a longer search in unknown territory will uncover a better target. There is no certainty a better target will be found with normal time and distance constraints. For most types of common crimes there are many more potential targets than an individual offender will victimize within their own activity space.

Basically people do not look in unfamiliar areas for common targets. The analogy is shoppers looking for bread or milk, they go to stores they know. Someone who is planning to steal from cars, looks at cars in their usual activity space. Why explore in an area where you do not even know what type of cars are there?

Crime occurrence space is within a high probability part of activity space that intersects with the location of targets seen as suitable to an offender. This is intuitively plausible as outside of their activity space, they will have to hunt for criminological opportunities and more likely face uncertain or unforeseeable risks.

7.1.3 Activity Space

Human cognition, spatial decision-making and human movements help to describe the activities of individuals—a way of thinking that has a long history in neurology, geography, and psychology [2, 3, 14, 21, 27, 33, 47]. People do not move randomly across urban landscapes [22]. For the most part, they commute between a handful

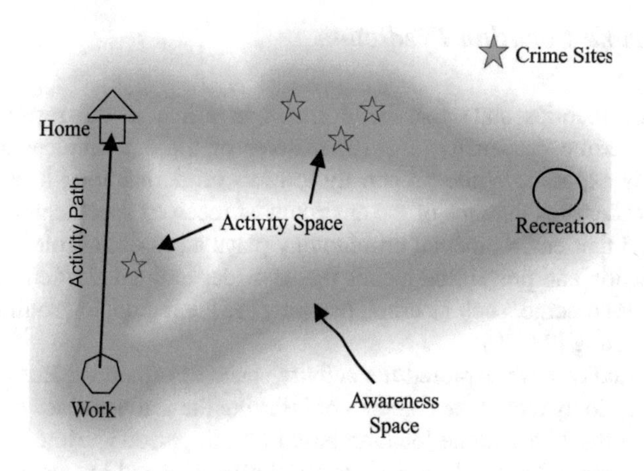

Fig. 7.1 Activity space

of routinely visited places like home, work, recreational facilities, and favorite shopping centers. With each and every trip, they will get more familiar with, and gain new knowledge about, these places and everything along the way. Eventually, a person will be at ease with a place. At this point, the place becomes part of the person's activity space, illustrated in Fig. 7.1. Activity space has two main components: Nodes and Paths [10]. The Nodes, called activity Nodes, are the locations that the person frequents, such as a workplace, residence, or recreation. These are the end-points of the journey. The Paths, called activity Paths, connect the Nodes and represent the person's path of travel between them.

7.1.4 Directionality

An offender commutes between his activity Nodes, but activity Paths that he chooses affect his activity and awareness spaces. Walkways and roadways that an offender selects for moving from one point to the next point in his commuting area are important in forming his activity and awareness spaces. On the other hand, the road network design and public transport system are very influential on formation of activity space of an offender.

Crime attractors and generators, locations where crime tends to cluster, affect offenders' directional preference. Crime generators are locations such as businesses, institutions that bring large numbers of people together. Crime attractors and generators affect directionality of offenders' movement [11, 17–19, 41].

7.1.5 Crime Location Prediction

Criminology theories state that involvement in crime is the result of: (1) an individual's crime propensity; and (2) features of the environment to which an individual is exposed. While propensity towards crime has long been studied, in the past few decades features of the environment received specific attention, and it is concluded that environmental criminology plays an essential role in developing crime reduction and prevention tactics that consider individual offenders [9]. New research areas emerge, such as crime mapping [25], geographic profiling [35], and crime forecasting [23, 30].

Several studies have explored the activity space of offenders. Rossmo [35] has developed a widely recognized method of inferring the activity space of an offender to determine the likely home location based on the person's crime locations. His approach is most often used for serious serial offenders and known as Geographic Profiling. He assumes that offenders will select targets and commit crimes near their home address or another major activity node or anchor point. Using this assumption, each new crime location is plotted on a map and a distance-decay function is used to calculate a probability space around each crime to denote the possible home location (and corresponding probability) of the offender. Geographic Profiling narrows down the probable home/nodal location of an offender more accurately with increasing number of crimes associated with the offender.

Canter [13] splits movement patterns of offenders into commuters and marauders. Marauders use a fixed base location (home, for example) and commit their crimes around it, making geographic profiling on this type of offender possible. According to Canter, and consistent with Crime Pattern Theory, marauders derive their offending locations from spatial patterns of their non-criminal daily activities. Although commuters probably also have a consistent base location, they travel to other places to commit crimes. Such travel patterns must be taken into account, making geographic profiling much more difficult.

Frank [20] proposed an approach to infer the activity Paths of all offenders in a region based on their crime and home locations. Assuming the home location as the center of an offender's movements, the orientation of Activity Paths of each individual offender was calculated so as to determine the major directions, relative to their home location, into which they tended to move to commit crimes.

Based on criminological theories, several studies propose mathematical models of spatial and temporal characteristics of crime to predict future crimes. However, these models do not predict individual offender behaviour. For instance in [30], the authors use a point-pattern-based transition density model for crime space-event prediction. This model computes the likelihood of a criminal incident occurring at a specified location based on previous incidents. In [40], the authors model the emergence and dynamics of crime hotspots. This work uses a two-dimensional lattice model for residential burglary, where each location is assigned a dynamic attractiveness value, and the behaviour of each offender is modeled with a random walk process. The authors study the impact of the model parameters on hotspot formation using a computer simulation.

Note that all of the above-mentioned methods solve related but different problems to which the experiments presented here cannot be compared. The model presented in [30] only predicts the time and location of the crime in the aggregate level. For a different purpose but similar to our work, Short et al. [40] uses standard random walk to model offenders' criminal behaviour. The method proposed in [35] and [13] discover offender home locations based on his crime locations. And finally, the output of the method proposed in [20] is locations which are centers of interest for committing crime. However, we compare to different *Collaborative Filtering* methods which are used for location recommendation in location-based social networks [49, 52]. Collaborating filtering (CF) infers the user's implicit preference form the explicit opinions of similar users based on the idea that users with similar behaviour in the past will have similar behaviour in the future [32].

7.1.6 Urban Environment

Intuitively, a road network can be decomposed into *road segments*, each of which starts and ends at an intersection. We use the *dual* representation where the role of roads and intersections is reversed. All physical locations along the same road segment are mapped to the same node. Formally, a road network is an undirected graph $R(L,Q)$, where L is a set of nodes, each representing a single road segment. Road segments l_j and l_k are connected, $\{l_j, l_k\} \in Q$, if they have an adjacent intersection in common. Crime locations within a studied geographic boundary are mapped to the closest road segment. Henceforth, the term "road" is used to refer to a road segment.

A vector $\bar{y}_j$ denotes the features of the road l_j including road length d_j, and *road attractiveness* features vector $\bar{a}_j$. Further, $\bar{a}_j$ is a vector of size m where the value of the kth entry of $\bar{a}_j$ corresponds to the total number of crimes of type k committed previously at l_j. Π_j denotes the set of neighbors of road l_j in the road network. $\Delta \subset L$ denotes a set of roads with the highest crime rate, called crime *hotspots*. D_{l_j,l_k} is the distance of road l_j from hotspot $l_k \in \Delta$, and f_j denotes the number of crimes at road l_j.

L_i is the set of roads at which offender u_i has been observed, including all of his known home and crime locations. $f_{i,j}$ and $t_{i,j}$, respectively, denote the frequency and the last time u_i was at anchor location l_j. *Offender trend* is given by a vector $\bar{x}_i$ of size m which indicates the crime trend of u_i as extracted from his criminal history. That is, the value of the kth entry of $\bar{x}_i$ corresponds to the number of crimes of type k committed by offender u_i.

7.1.7 Problem Definition

Given a crime dataset $\mathscr{C}$, an offender u_i and road network $R(L,Q)$ associated with $\mathscr{C}$, the goal is to learn the *activity space distribution* F for u_i on R [43–45]. That is, for each road $l_j \in L$, $F(i,j)$ states the probability that l_j is part of the activity space of u_i, and thus the likelihood for offender u_i committing a crime at road l_j is

$$F(i,j) \longrightarrow [0,1] \text{ with } \sum_{j=1}^{|L|} F(i,j) = 1 \tag{7.1}$$

By learning the activity space distribution of individual criminal offenders, we obtain a probabilistic model of offender activity space that can be used for personalized prediction of future crime locations of the offender. The assumption is that the richer and more detailed the offender profile is, the more accurate is the probabilistic activity space model, and also the prediction of future crime locations. This probabilistic view of activity space means that there is no sharp boundary between activity space and awareness space, which directly corresponds to the intuitive understanding of the concept of activity space in criminology.

7.2 CrimeTracer Model

In this section, we present CrimeTracer [43–45], our proposed crime spatial analysis model.

7.2.1 Model Description

A *random walk* over a graph is a stochastic process in which the initial state is known and the next state is decided using a transition probability matrix that identifies the probability of moving from a node to another node of the graph. Under certain conditions the random walk process converges to a stationary distribution [24], which assigns an importance value to each node of the graph.

The random walk method satisfies the locality aspect of crimes, which states that offenders do not attempt to move far from their anchor locations. But it has some shortcomings that we aim to address in the CrimeTracer model.

The CrimeTracer model consists of three important components: an offender, the road network, including all locations where the offender committed crime, and the co-offending network that connects offenders. Starting from an anchor location, the offender explores the city through the underlying road network. At each road he decides whether to proceed to a neighboring road or return to one of his anchor locations. The random walk process continues until it converges to the steady state which reflects the probability of visiting a road by the offender. This probability can be relevant to the offender's exposure to a crime opportunity.

For learning the activity space of an offender we need to understand his daily life and routines, but in the crime dataset generally we miss the Paths completely and the Nodes partially (refer to Sect. 7.1.3), which is a major challenge. To address these challenges, we improve our incomplete knowledge about offenders with the available information in the dataset. The set of anchor locations of each offender is extended by adding his co-offenders' anchor locations. The extended set is denoted as *main anchor locations*. This extension is motivated by the assumption that friends in the co-offending network are likely to share the same location.

For each offender, using a Gaussian model, we define his *intermediate anchor locations* as the roads closest to the set of his main anchor locations. An offender starts his random walk either from a main anchor location or from an intermediate anchor location.

Given that the actual trajectories in an offender's journey to crime are not known, the model guides offender movements in directions with a higher chance of committing a crime. This is done by taking into account two different aspects that influence offender movement directionality in computing the transition probabilities in a random walk. The first aspect refers to road characteristics in terms of road feature values: the number of crimes committed on this road for each different crime category and the road length. The second aspect refers to the personal preferences of each offender for certain types of crime, as stated in the offender profile, as a driving factor in the decision process when encountering a crime opportunity. Whenever none of the neighbors of the current road promise any crime opportunities of interest, road length is the single determining factor. Using the *supervised random walks* method [1], we learn the importance of these features and exploit them in computing the transition probability matrix for the random walk.

The second proposed approach for learning the movement directionality of an offender uses the concept of crime generators and attractors [11]. These are two types of locations where crimes tend to cluster. Assuming that offenders are drawn in directions leading toward criminal attractors, we assign a higher probability to roads leading toward crime hotspots—much like gravity centers affecting the random walk.

The random walk stops in a road which provides an opportunity for committing a crime, depending on both road characteristics and offender crime preferences. Below we describe different elements of the proposed model in detail.

7.2.2 *Random Walk Process*

For each single offender, we perform a series of random walks on the road network $R(L,Q)$. In each random walk the offender starts his exploration from one of his anchor locations, traversing the road network to locate a criminal opportunity.

For offender u_i, the random walk process starts from one of his anchor locations with predefined probabilities as described in Sect. 7.2.3. At each step k of the random walk, the offender is at a certain road l_j and makes one of two possible decisions:

- with probability α he decides to return to an anchor location and not look for a criminal opportunity this time, choosing an anchor location in one of two ways:
  - with probability β he decides to return to a main anchor location $l \in \mathcal{L}_i$.
  - with probability $1-\beta$ he returns to an intermediate anchor location $l \in \mathcal{I}_i$.
- with probability $1-\alpha$ he continues looking for a crime opportunity.

If he continues his random walk, then he has two options in each step of the walk:

- with probability $\theta(u_i, l_j, k)$ stop the random walk, which means the offender commits a crime at road l_j.
- with probability $1-\theta(u_i, l_j, k)$ continue the random walk, moving to another road which is a direct neighbor of l_j.

To continue the random walk at road l_j, we select a direct neighbor road from Π_{l_j}. The probability of selecting road segment l_k in the next step is defined as

$$P(l_j \to l_k) = \frac{\phi_{\bar{w}}(l_k)}{\sum\limits_{l_p \in \pi_{l_j}} \phi_{\bar{w}}(l_p)} \tag{7.2}$$

The probability of being at road l_r at step $k+1$ given that the offender was at road l_j at step k is

$$\begin{aligned} P(X_{l,k+1} = l_r | X_{l,k} = l_j) &= (1-\alpha)(1-\theta_{l_j,k}) \times P(l_j \to l_r) \\ &= (1-\alpha)(1-\theta_{l_j,k}) \times \frac{\phi_{\bar{w}}(l_r)}{\sum\limits_{l_s \in \Pi_{l_j}} \phi_{\bar{w}}(l_s)} \end{aligned} \tag{7.3}$$

We terminate the random walks when $||F^{m+1}|| - ||F^m|| \leq \varepsilon$, where $F^m = \begin{pmatrix} F(u_i, l_1) \\ \vdots \\ F(u_i, l_{|L|}) \end{pmatrix}$ is the results for u_i after m random walks. For some offenders the random walks do not converge, in which case we terminate the overall process at $m > 10{,}000$.

7.2.3 Starting Probabilities

CRIMETRACER distinguishes two different types of starting nodes:

- Main anchor locations are all anchor locations of a single offender and his co-offenders: $\mathscr{L}_i = L_i \cup \{l_j : l_j \in L_v, v \in \Gamma_u\}$. Co-offending links are important since they are the reasons for many spatial effects related to crime [36]. It is concluded that offenders who are socially close are spatially close too [42]. The rationale is that offenders who have collaborated in the past likely may have shared information on anchor locations in their activity space, an aspect that possibly affects their choice of future crime locations. In computing the starting probability of each anchor location, the two primary factors are the frequency and the last time an offender visited an anchor location. The probability that offender u_i starts his random walk from l_j thus is

$$S(i,j) = \frac{f_{i,j} \times e^{\frac{-(t-t_{i,j})}{\rho}}}{\sum_{l_k \in \mathscr{L}_i} f_{i,k} \times e^{\frac{-(t-t_{i,k})}{\rho}}} \tag{7.4}$$

where t is the current time, and ρ is the parameter controlling the effect of the timing.
- Intermediate anchor locations are the closest locations to main anchor locations. Human mobility models use Gaussian distribution to analyze human movement around a particular point such as home or work location [12, 22]. We assume that offender movement around his main anchor locations follows a Gaussian distribution. Each main anchor location of offender u_i is used as the center, and the probability of u_i being located in a road is modeled with a Gaussian distribution. Given road l the probability of u_i residing at l is computed as follows:

$$S(i,l) = \sum_{l_j \in \mathscr{L}_i} \frac{f_{i,j}}{\sum_{l_k \in \mathscr{L}_i} f_{i,k}} \frac{\mathscr{N}(l|\mu_{l_j}, \Sigma_{l_j})}{\sum_{l_k \in \mathscr{L}_i} \mathscr{N}(l|\mu_{l_k}, \Sigma_{l_k})} \tag{7.5}$$

Here l is a road which does not belong to the set of main anchor locations. $\mathscr{N}(l|\mu_{l_j}, \Sigma_{l_j})$ is a Gaussian distribution for visiting a road when u_i is at anchor location l_j, with μ_{l_j} and Σ_{l_j} as mean and covariance. We consider the normalized activity frequency of u_i at l_j, meaning that a main anchor location with higher activity frequency has higher importance. For offender u_i, the roads with the highest probability of being an intermediate anchor location are added to the set $\mathscr{I}_i$ as additional starting nodes besides the main anchor locations.

7.2.4 Movement Directionality

As discussed in Sect. 7.1.4, directionality of offender movement plays an important role in activity space formation. We propose here two approaches to determining movement directionality. The first approach learns the weights of the features that determine the probability of selecting a road among all neighbor roads in a random walk process. The second approach leads an offender in the direction that gets him closer to the crime hotspots.

Hotspots Influence In this approach the transition probability is computed based on proximity of a road to the crime hotspots and the importance of each crime hotspot, which is proportional to the number of crimes committed there. The function $\phi(l_j)$ is used in computing the transition probability (refer to Sect. 7.2.2) of moving offender u_i from l_k to l_j:

$$\phi(l_j) = \sum_{n=1}^{|\Delta|} D_{k,n} \times f_n \tag{7.6}$$

where $D_{k,n}$ is the distance of road l_k from the hotspot $l_n \in \Delta$, which is equal to the length of shortest path between two roads on the road network. f_n is the number of crimes committed at l_n.

Learning Road Feature Weights Road feature weights $\bar{w}$ are used to compute the transition probabilities. The function $\phi_{\bar{w}}(l_j)$ is computed based on the road features

$$\phi_{\bar{w}}(l_j) = \sum_{k=1}^{m+1} w_k \times y_{j,k} \tag{7.7}$$

where $\bar{y}_{j,k}$ is the value of kth feature of the road l_j, and w_k is the corresponding weight of the feature k.

We use the same idea used in the supervised random walks method [1] for link prediction in social networks. This method guides the random walk toward the preferred target nodes by utilizing node and edge attributes.

Each offender in a random walk starting from his home location reaches a crime location. In the training data for each offender we have a series of crime journeys, meaning that for a source node s we have a set of destination nodes $D = \{d_1, d_2, \ldots, d_n\}$, and a set of non-destination nodes $Z = \{z_1, z_2, \ldots, z_m\}$. The probability of visiting a node p_d is influenced by the road transition probabilities. And the transition probabilities are dependent on the road feature weights. Now, we say an offender starting from node s so as to visit destination nodes $d_i \in D$ more often than non-destination nodes $z_i \in Z$ by formulating the following optimization problem:

$$\min_{\bar{w}} F(\bar{w}) = \|\bar{w}\|^2 + \lambda \sum_{d \in D, z \in Z} \text{loss}(p_z - p_d) \tag{7.8}$$

where λ is the regularization parameter, and loss is a predefined loss function for penalizing the cases in which the stationary probability of a non-destination node p_z is higher than the stationary probability of a destination node p_d.

7.2.5 Stopping Criteria

The probability of stopping the random walk for an offender at a given road corresponds to the probability of this offender committing a crime in that road segment. Two factors influence the stopping probability of offender u_i in the road l_j. The first one relates to the similarity of the crime trend of offender u_i and the criminal attractiveness of road l_j, where higher similarity means a higher chance that u_i's random walk stops at l_j. The second factor is the distance of l_j from the starting point measured in the number of steps from the starting point. To satisfy the locality aspect of crimes, the probability of continuing the random walk should decrease while getting farther from the starting point:

$$\theta(u_i, l_j, k) = \mathrm{Sim}(i,j) \times \frac{1}{1+e^{\frac{-k}{2}}} \tag{7.9}$$

where $\mathrm{Sim}(i,j)$ denotes the cosine similarity of crime trend of u_i and the road attractiveness of the road l_j. The stopping probability is inversely proportional to the step number k:

$$\mathrm{Sim}(i,j) = \frac{\bar{x}_i.\bar{a}_j}{|\bar{x}_i||\bar{a}_j|} \tag{7.10}$$

7.3 Experiments and Results

In this section, we present our experimental design, the comparison partners, and the results [43–45].

7.3.1 Data Characteristics

For the study presented here, we concentrate on the use of a subset of this dataset which includes all crimes in Metro Vancouver, B.C. (total population: over 2.4 million), where different regions are connected through a road network composed of 64,108 road segments with an average length 0.2 km. Table 7.1 shows a statistics for the used crime dataset. Figure 7.2 shows the spatial distribution of crimes in Metro Vancouver.

Table 7.1 Statistical properties of the Metro Vancouver crime dataset

Property	Value
Number of crimes	125,927
Number of offenders	189,675
Number of offenders with more than one crime	25,162
Offenders with more than one crime location	18,615
Number of co-offending links	68,577
Number of co-offenders in co-offending network	17,181
Average node degree in co-offending network	4
Number of road segments	64,108
Average crime per road segment	2

Fig. 7.2 Crime distribution in Metro Vancouver; *red dots* show crime locations, and *black lines* show the major roads

Figure 7.3a and b illustrate the distribution function of crime incidents per offender and per road segment. Both distributions have heavy-tailed pattern. 83 % of the offenders committed only one crime, while less than 1 % of the offenders committed ten or more crimes. Further, 38 % of the road segments are linked to at least one crime and 9 % are linked to ten or more crimes. Half of all the crimes occurred in only 1 % of all road segments, and a total of 25 % in only 100 road segments.

Figure 7.4a and b, respectively, show the average home location to crime location distance and the average distance between crime locations for all offenders in the dataset. The average home to crime location distance of 80, 63, and 40 % of all offenders is less than 10, 5, and 2 km, respectively. And the average crime location distance of 73, 52 and 26 % of all offenders is less than 10, 5, and 2 km, respectively. One can assume that frequent offenders are generally mobile and have several home locations identified in their records. In fact, 41 % of the offenders who committed more than one crime have more than one home location.

The dataset differentiates more than 1,000 crime types, with half of them occurring only a few times. For three well-defined categories of personal crime

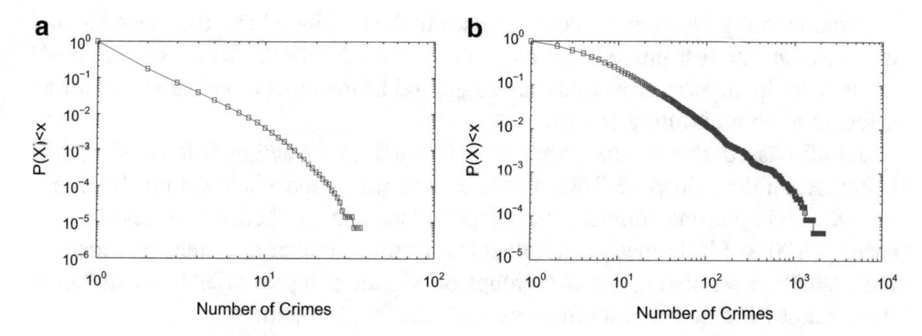

Fig. 7.3 Distribution function: (a) Crimes per offender; (b) Crimes per road segment

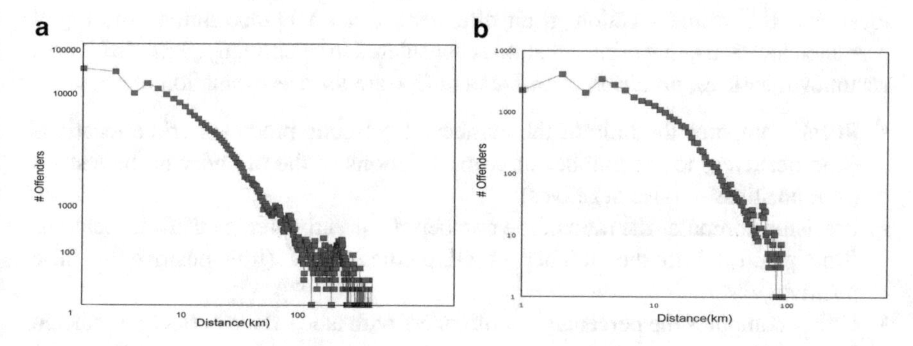

Fig. 7.4 Avg. distance (a) home–crime locations; (b) crime–crime locations

(like assault), property crime (like break & enter), and drug crime, as expected, the property crime category has the largest average home location to crime location distance. For half of the repeat offenders, at least half of their crimes belong to only one category, meaning that half of the repeat offenders specialize in at least one category, and they keep their crime trend for a while.

7.3.2 Experimental Design

For each offender we order his crime events chronologically based on their time. Then we split these events into a training set and a test set. The first 80 % of the crimes are used for training the model which predicts the offender activity space. The remaining 20 % of crimes are used for testing the model. We consider only offenders with at least two different crimes which includes about 10 % of the offenders in the crime dataset. We note that the training data used for learning road features as described in Sect. 7.2.4 is not included in the evaluation to prevent biasing CRIMETRACER.

After learning the offender activity space in the training phase, the trained model is applied in the test phase to predict future crime locations. To do so, the top-N roads with the highest probability are suggested as the most probable places for an offender to commit future crimes.

As discussed above, the focus of this work is modeling offenders' spatial behaviour in the coldspots. Thus, in our experiments we exclude the top 100 roads with the highest crime numbers, the hotspots. The number of crimes in these hotspot roads is 100 to 1100 times greater than the average number of crimes in a road. In the evaluation we distinguish two groups of offenders: repeat offenders with ten or more crimes and non-repeat offenders with less than ten crimes.

To evaluate the accuracy of activity space prediction, we measure the number of crimes committed by an offender in his testing dataset among the top-N predicted locations. If a crime location in an offender's test set is also among the top-N predicted locations, that crime location is considered to be correctly *predicted*. Three accuracy measures, *precision*, *recall*, and *utility*, are used as evaluation metrics:

- Recall computes the ratio of the number of correctly predicted crime locations (true positives) to the number of crime locations of the offender in the test set (true positives + false negatives).
- Precision computes the ratio of the number of correctly predicted crime locations (true positives) to the number of all predictions N (true positives + false positives).
- Utility computes the percentage of offenders with at least one correctly predicted crime location.

Recall and precision are averaged across all offenders to determine the overall performance for different values of N. In computing the precision value for an offender, if the activity space contains $M < N$ roads, we use M instead of N.

7.3.3 Comparison Partners

In this section, we introduce different versions of CrimeTracer and the comparison partners methods used in our performance evaluation.

For evaluating the CrimeTracer performance, we test the two different movement directionality approaches and the following types of locations included in the activity space of offenders. For every offender locations are categorized into three groups: (a) *Known locations* that includes home and crime locations of the offender. (b) *Derived locations* which are locations shared with co-offenders and intermediate anchor locations. These locations are derived from observed information in the crime dataset. (c) *Unknown locations* that includes any location which is not a known or derived location.

For a deeper understanding of CrimeTracer performance and the role of each of the above-mentioned location types, we consider three approaches : (1) In the first approach (denoted by U) we include only unknown locations in the activity space of an offender and consequently in the crime location prediction; (2) In the second approach (denoted by D) we include only unknown and derived locations in the activity space of an offender; and (3) In the last approach (denoted by A) all locations are considered.

Two different movement directionality methods are introduced in Sect. 7.2.4: hotspot influence (denoted by H) and learning road feature weights (denoted by F). For each of these CRIMETRACER versions we consider the three above-mentioned evaluation approaches. For instance, CRIMETRACER-HU denotes CRIMETRACER using the hotspot influence method (H) for movement directionality that includes only unknown locations (U) in the predicted locations.

As discussed in Sect. 7.1.5, there is no related work that solves the problem of personalized crime location prediction. However we use the following methods which are equivalent to state-of-the-art methods for location recommendation [49]:

Random Walk This is the standard random walk with restart method (RWR) [46].

Hotspots Using the basic hotspot approach (HS), roads are ranked based on the number of crimes in that road.

Proximity In the proximity approach (DS) we rank the roads based on their distance from the offender's anchor locations. Here distance denotes the length of the shortest path between two roads on the road network.

Offender-Based CF The intuition behind the offender-based CF approach (OCF) is that offenders who had similar behaviour in the past will have similar behaviour in the future. Let $b_{ij} = 1$ if $l_j \in \mathscr{L}_i$, and $b_{ij} = 0$ if $l_j \notin \mathscr{L}_i$. Now $F(i,j)$ is the probability of a crime committed in road l_j by u_i:

$$F(i,j) = \frac{\sum\limits_{u_k \in V \wedge k \neq i} \mathrm{Sim}(i,k).b_{k,j}}{\sum\limits_{u_k \in V \wedge k \neq i} \mathrm{Sim}(i,k)} \tag{7.11}$$

where $\mathrm{Sim}(i,k)$ denotes the cosine similarity measure between offenders u_i and u_k:

$$\mathrm{Sim}(i,k) = \frac{\sum\limits_{l_j \in L} b_{i,j}.b_{k,j}}{\sqrt{\sum\limits_{l_j \in L} b_{i,j}^2}\sqrt{\sum\limits_{l_j \in L} b_{k,j}^2}} \tag{7.12}$$

Location-Based CF In location-based CF (LCF) we consider the similarity of locations instead of the similarity of offenders:

$$F(i,j) = \frac{\sum\limits_{l_k \in L \wedge k \neq j} \mathrm{Sim}(j,k).b_{i,k}}{\sum\limits_{l_k \in L \wedge k \neq j} \mathrm{Sim}(j,k)} \tag{7.13}$$

where $\mathrm{Sim}(j,k)$ is the cosine similarity measure between roads l_j and l_k:

$$\mathrm{Sim}(j,k) = \frac{\sum_{u_i \in V} b_{i,j}.b_{i,k}}{\sqrt{\sum_{u_i \in V} b_{i,j}^2}\sqrt{\sum_{u_i \in V} b_{i,k}^2}} \tag{7.14}$$

Co-offending-Based CF Co-offenders can share their information about criminal opportunities and take advantage of this information in committing a new crime. Co-offending-based CF (SCF) computes the probability of a crime being committed in road l_j by u_i as follows:

$$F(i,j) = \frac{\sum_{u_k \in \Gamma_i} \mathrm{Sim}(i,k).b_{k,j}}{\sum_{u_k \in \Gamma_i} \mathrm{Sim}(i,k)} \tag{7.15}$$

$\mathrm{Sim}(i,k)$ denotes the geo-social influence between u_i and u_k and is defined as follows:

$$\mathrm{Sim}(i,k) = \frac{|\Gamma_i \cap \Gamma_k|}{|\Gamma_i \cup \Gamma_k|} + \frac{|\mathcal{L}_i \cap \mathcal{L}_k|}{|\mathcal{L}_i \cup \mathcal{L}_k|} \tag{7.16}$$

7.3.4 Experiments and Results

CRIMETRACER Scenarios Figures 7.5, 7.6, and 7.7 show performance of six different versions of CRIMETRACER including CRIMETRACER-HU CRIMETRACER-HD, CRIMETRACER-HA, CRIMETRACER-FU, CRIMETRACER-FD, and CRIMETRACER-FA in terms of recall, precision, and utility measures.

With regard to the type of locations included in the prediction process, as expected CRIMETRACER-HA and CRIMETRACER-FA have the best performance, and CRIMETRACER-HU, and CRIMETRACER-FU have the worst performance. The recall of CRIMETRACER-HA, CRIMETRACER-HD and CRIMETRACER-HU for $N = 20$ is 23.4, 10.2, and 5.9 %, respectively. Derived and known locations increase the recall by 4.3 and 13.1 %, respectively. We observe a similar result when comparing the performance of CRIMETRACER-FA, CRIMETRACER-FD, and CRIMETRACER-FU.

An important question is which of these scenarios should be used in a real-world application of CRIMETRACER. According to criminological theories such as exact-repeat/near repeat event [28] and broken window theory [51], known locations of offenders are always likely places to commit a new crime. The results presented in this section also support this idea. In a real-world application known locations may be included in the predicted locations automatically. One may conclude that CRIMETRACER-HD and CRIMETRACER-FD are more appropriate versions of CRIMETRACER for a real-world application.

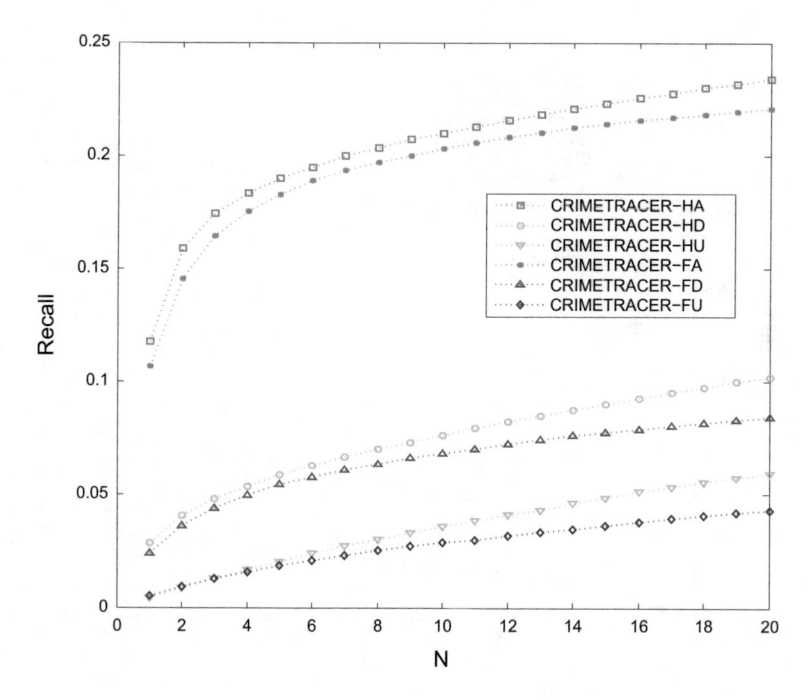

Fig. 7.5 Recall of different versions of CRIMETRACER for different values of N

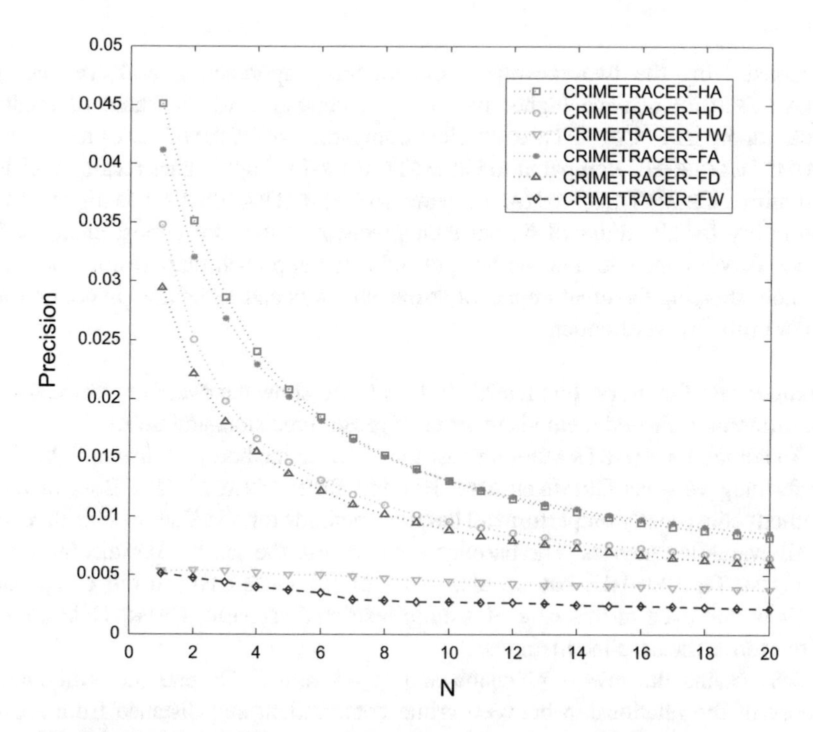

Fig. 7.6 Precision of different versions of CRIMETRACER for different values of N

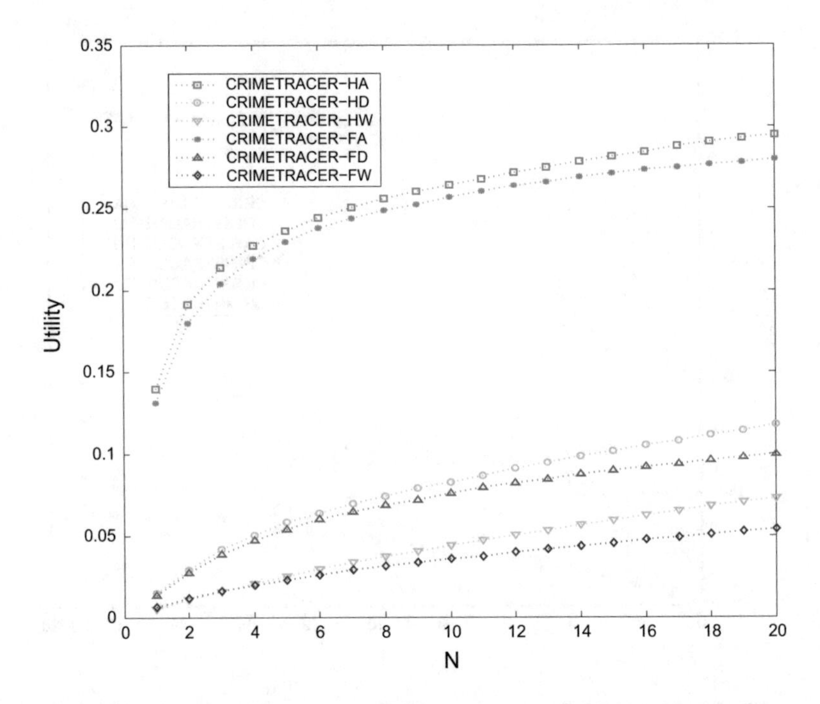

Fig. 7.7 Utility of different versions of CrimeTracer for different values of N

Considering the two movement directionality approaches, both versions of CrimeTracer achieve higher performance compared to the standard random walk approach. CrimeTracer-HD compared to CrimeTracer-FD and CrimeTracer-HU compared to CrimeTracer-FU have higher recall, precision and utility. CrimeTracer-HA compared to CrimeTracer-FA has higher recall and utility for all values of N, but their precision values are almost identical for $N >= 6$. We conclude that the hotspot influence approach outperforms the other method, showing the great impact of crime attractors and generators in committing a new crime by an offender.

Comparison Partners Figures 7.8, 7.9, and 7.10 show the overall performance of the different evaluated methods in terms of recall, precision, and utility.

To compare CrimeTracer against the baseline methods, we use only the best performing versions CrimeTracer-HD and CrimeTracer-HU. Both of these methods consistently outperform all baseline methods for all values of N with regard to all evaluation metrics. The baseline methods use the same experimental design as CrimeTracer-HD, but we also test CrimeTracer-HU in the comparison to show that even in this case of a more restricted scenario, CrimeTracer still outperforms the baseline methods.

DS obtains the lowest precision and recall values. Despite the well-studied theory of the relationship between crime commitment and distance from anchor

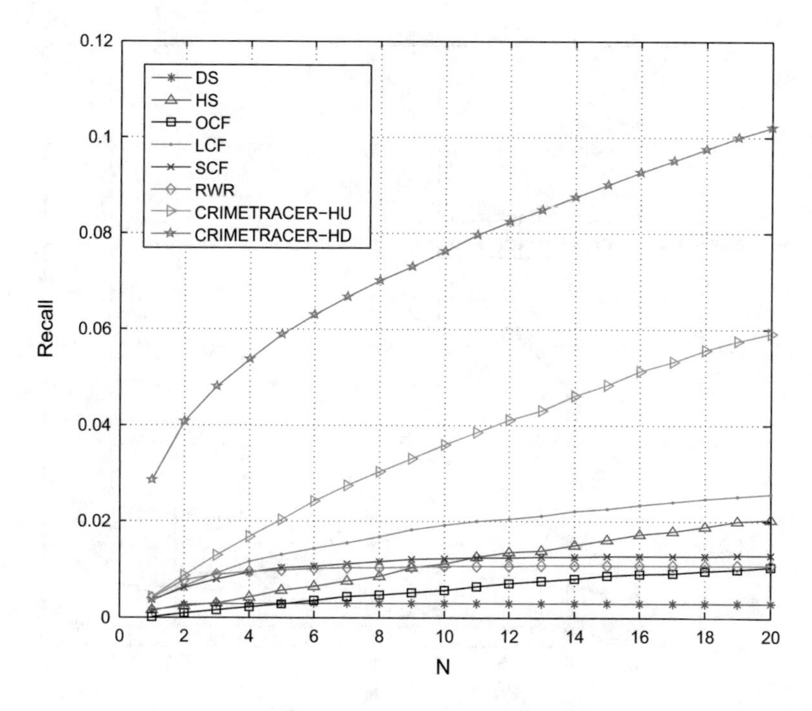

Fig. 7.8 Recall for different values of *N*

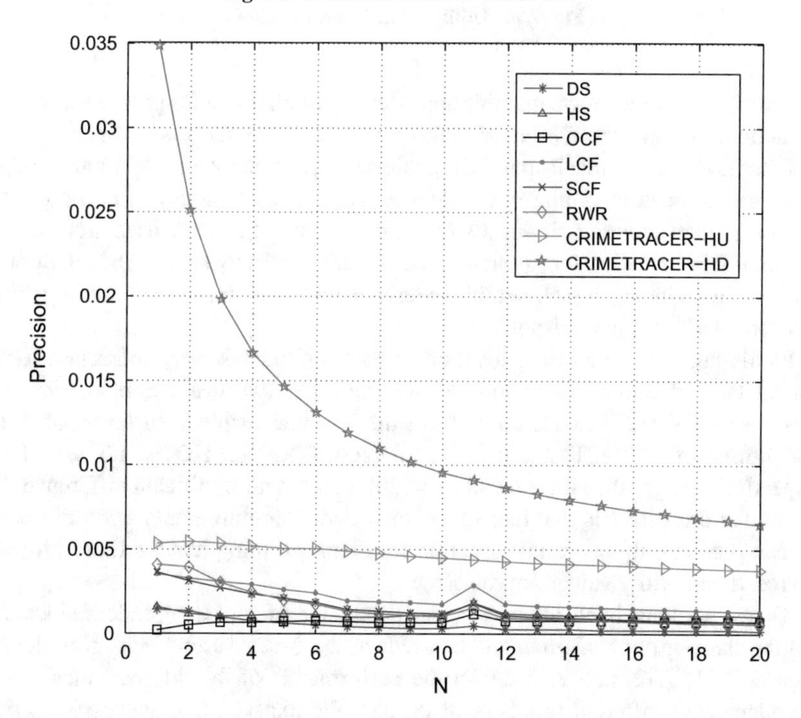

Fig. 7.9 Precision for different values of *N*

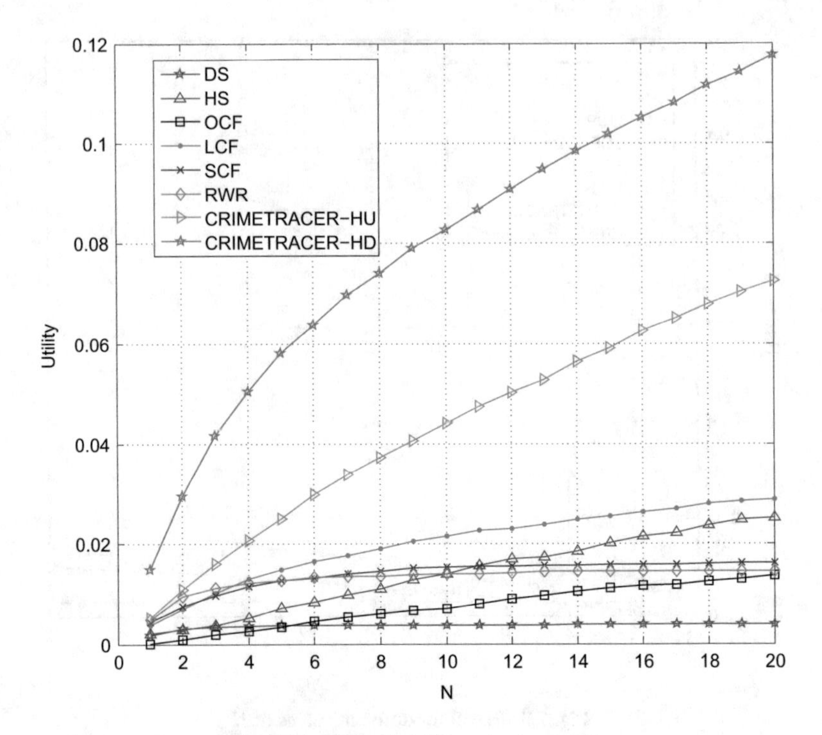

Fig. 7.10 Utility for different values of N

locations, this result shows that this approach is not effective for personalized crime prediction. Among the CF-based approaches, OCF has the poorest performance. LCF achieves better recall, but SCF achieves higher precision. It is interesting to observe that location similarity contributes more to the accuracy of crime location prediction than offender similarity. One can conclude that SCF uses more reliable but limited information for predicting the offenders activity space. The recall of HS improves with increasing N, but this method naturally is strong in predicting crimes in hotspots and not in coldspots.

Predicting even one crime location of each offender is very important for the critical task of crime prevention. As for the other two evaluation metrics, both versions of CRIMETRACER outperform the baseline methods in terms of utility. The utility of CRIMETRACER-HU and CRIMETRACER-HD is 1.3 and 1.5 %, respectively, larger than their recall ($N = 20$), making no significant difference. One reason for this effect is that half of the offenders committed only two crimes, and we can predict only one crime location for them, meaning that for these offenders the recall and utility values are the same.

There has long been interest in the behaviour of repeat offenders since controlling these groups of offenders can reduce the overall crime level significantly. Figures 7.11, 7.12, and 7.13 depict the performance of the different methods for offenders with different numbers of crimes. We expect more successful activity

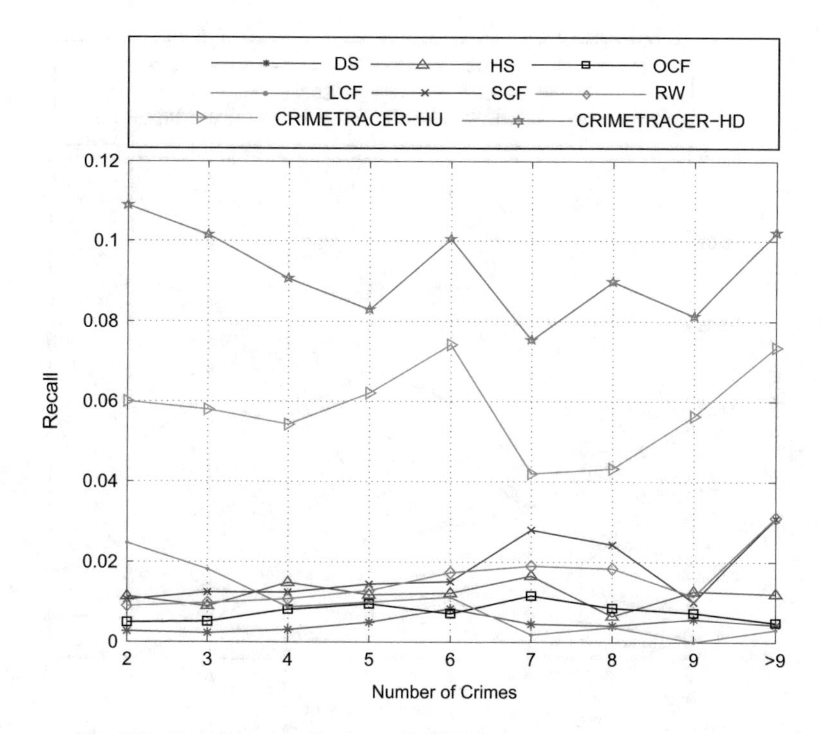

Fig. 7.11 Recall for the offenders with different number of crimes ($N = 20$)

space learning for offenders who have committed more crimes, and for whom we have more information. We observe such a trend for CRIMETRACER-FU, where the average recall for offenders who committed only two crimes is about 4 % while this value increases by 3 % for offenders who committed ten or more crimes, as well as for RWR and SCF.

Interestingly the hotspot influence approach causes a significant increase in recall of non-repeat offenders (the biggest group of offenders). Comparing CRIMETRACER-HU to CRIMETRACER-FU, the recall increases by 2 % for this group of offenders, while the recall for repeat offenders is almost equal for these two methods. On the other hand, while for CRIMETRACER-FU the recall of repeat offenders is 3 % higher than the recall of non-repeat offenders, this difference is only 1 % for CRIMETRACER-HU. Thus, the directionality movement approach influenced by hotspot locations contributes more to the recall of non-repeat offenders than to the recall of repeat offenders.

While we do not observe a significant increase in recall of repeat offenders compared to non-repeat offenders for either of the CRIMETRACER versions, we observe such a trend in the precision measure. Another interesting observation is that for SCF using co-offending information causes a significant performance gain for repeat offenders who have higher co-offending rates.

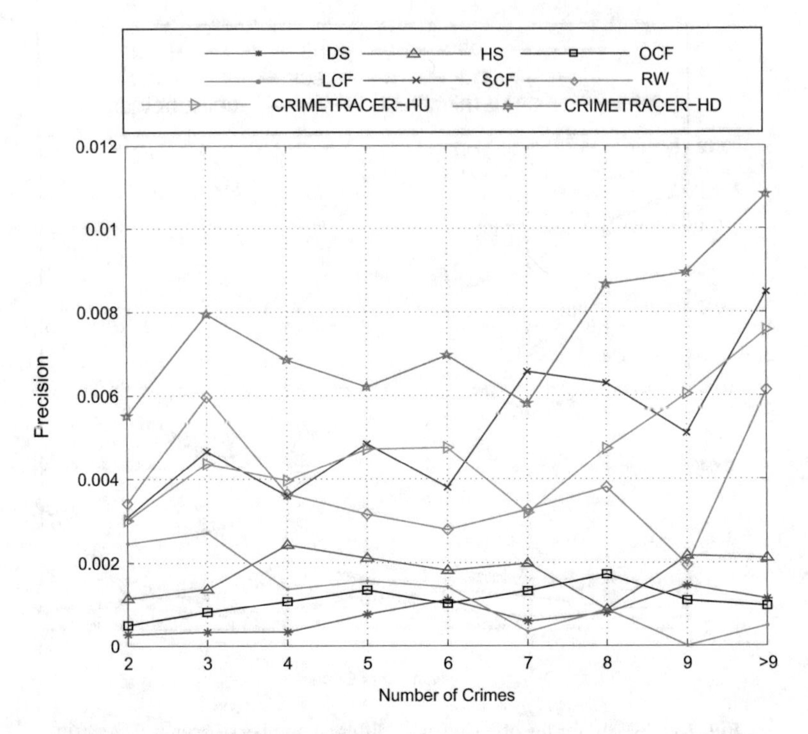

Fig. 7.12 Precision for the offenders with different number of crimes ($N = 20$)

Non-repeat offenders are the majority of offenders, and in this study half of the offenders used for the evaluation committed only two crimes. As shown in Figs. 7.11, 7.12, and 7.13, for non-repeat offenders CRIMETRACER-HU and CRIMETRACER-HD outperform the baseline methods by large margins. We notice that LCF also works well for offenders who committed only two crimes. This interesting result shows that beginner offenders tend to commit crimes in common locations. On the other hand, while SCF is not accurate for beginners, with increasing crime numbers its performance increases significantly. This means that being more experienced in crime boosts the number of co-offenders and consequently the chance of sharing criminal opportunities.

CRIMETRACER Elements We studied the contribution of different components of CRIMETRACER to its performance. Compared to the standard random walk with restart, CRIMETRACER incorporates additional anchor locations (co-offending information and intermediate anchor locations), movement directionality, and stopping criteria. We added these components separately to RWR to determine their individual contribution. Table 7.2 shows the results. The strongest component is the stopping criteria and the weakest is the learning of road feature weights. The main idea behind the stopping criteria is to stop the random walk of an offender in a road where the crime history is similar to the offender crime trend. However combining

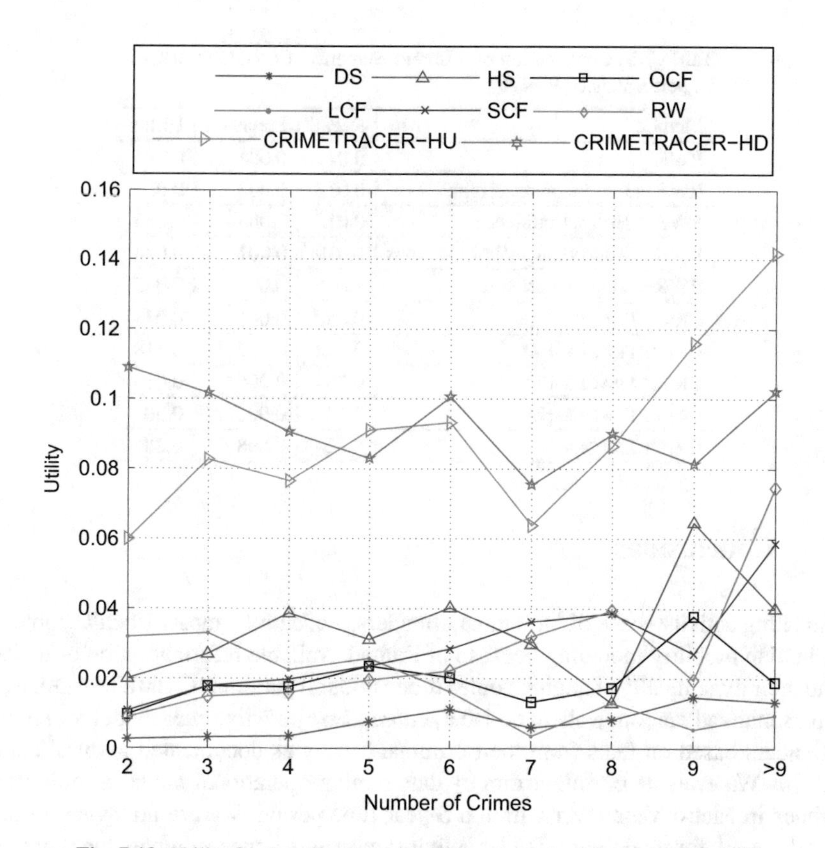

Fig. 7.13 Utility for the offenders with different number of crimes ($N = 20$)

all components in CrimeTracer-HD achieves the best result and improves the performance of RWR significantly in terms of all evaluation metrics. We include the performance of other versions of CrimeTracer in Table 7.2 to be able to compare the performance of different versions of CrimeTracer more exactly.

We note that the overall performance of CrimeTracer is comparable to the performance of state-of-the-art methods for location recommendation [31, 49], where the information about users' spatial patterns is much denser than the available information about offenders. One may criticize that in location recommendation the exact locations are predicted while in CrimeTracer only roads are predicted as offender activity space. However, as discussed in [15], roads are the natural domain for many policing activities, and a more realistic urban element for predicting a crime than the exact latitude and longitude. In addition, the road network we use in our study is in the micro scale with the average road length of 0.2 km.

Table 7.2 Contribution of different elements of CrimeTracer to its performance ($N = 20$)

Method	Recall	Precision	Utility
RWR	0.011	0.004	0.014
RWR + Road feature weights	0.013	0.003	0.017
RWR + Hotspot influence	0.015	0.003	0.016
RWR + Additional anchor locations	0.019	0.001	0.024
RWR + Stopping criteria	0.036	0.003	0.045
CrimeTracer-HU	0.059	0.006	0.073
CrimeTracer-HD	0.102	0.007	0.118
CrimeTracer-FD	0.084	0.006	0.010
CrimeTracer-HA	0.23	0.008	0.30
CrimeTracer-FA	0.22	0.008	0.28

7.4 Conclusions

Modeling activity space of individual offenders is one of the most difficult problems in human mobility modeling because of limited available information on offenders and their dynamically changing complex behavioural patterns. CrimeTracer uses a personalized random walk to derive a *probabilistic activity space* model for known offenders based on facts from their criminal history as documented in an offender profile. We evaluate our algorithm by data mining operational police records from crimes in Metro Vancouver within a 5-year time period. We are not aware of any similar work for modeling offender activity space and, hence, compare the proposed approach with location recommendation methods. CrimeTracer outperforms all other evaluated methods tested here. It boosts the prediction performance of the repeat offenders, compared to the non-repeat offenders, by using co-offending information. As expected, the chance of having co-offending links is higher for repeat offenders. All elements used in CrimeTracer, which are additional to the standard random walk model, contribute to the performance of this method.

References

1. L. Backstrom, J. Leskovec, Supervised random walks: predicting and recommending links in social networks, in *Proceedings of the Fourth ACM International Conference on Web Search and Data Mining (WSDM'11)*, 2011, pp. 635–644
2. W. Bernasco, Co-offending and the choice of target areas in burglary. J. Investig. Psychol. Offender Profiling **3**(3), 139–155 (2006)
3. W. Bernasco, Them again? Same-offender involvement in repeat and near repeat burglaries. Eur. J. Criminol. **5**(4), 411–431 (2008)
4. W. Bernasco, Modeling micro-level crime location choice: Application of the discrete choice framework to crime at places. J. Quant. Criminol. **26**(1), 113–138 (2010)
5. W. Bernasco, R. Block, Where offenders choose to attack: a discrete choice model of robberies in chicago. Criminology **47**(1), 93–130 (2009)

6. W. Bernasco, P. Nieuwbeerta, How do residential burglars select target areas? A new approach to the analysis of criminal location choice. Br. J. Criminol. **45**(3), 296–315 (2005)
7. D. Birks, M. Townsley, A. Stewart, Generative explanations of crime: using simulation to test criminological theory. Criminology **50**(1), 221–254 (2012)
8. A.A. Braga, The effects of hot spots policing on crime. Ann. Am. Acad. Pol. Soc. Sci. **578**(1), 104–125 (2001)
9. P.J. Brantingham, P.L. Brantingham, *Environmental Criminology* (Sage Publications, Beverly Hills, CA, 1981)
10. P.L. Brantingham, P.J. Brantingham, Nodes, paths and edges: considerations on the complexity of crime and the physical environment. J. Environ. Psychol. **13**(1), 3–28 (1993)
11. P.L. Brantingham, P.J. Brantingham, Criminality of place. Eur. J. Crim. Pol. Res. **3**(3), 5–26 (1995)
12. D. Brockmann, L. Hufnagel, T. Geisel, The scaling laws of human travel. Nature **439**(7075), 462–465 (2006)
13. D.V. Canter, A. Gregory, Identifying the residential location of rapists. J. Forensic Sci. Soc. **34**(3), 169–175 (1994)
14. E.R. Chrastil, Neural evidence supports a novel framework for spatial navigation. Psychon. Bull. Rev. **20**(2), 208–227 (2013)
15. T.P. Davies, S.R. Bishop, Modelling patterns of burglary on street networks. Crime Sci. **2**(1), Article 10, (2013)
16. M. Felson, R. Clarke, Opportunity makes the thief: practical theory form crime prevention. *Police Research Series*, Article 98 (Home Office, Policing and Reducing Crime Unit, Research, Development and Statistics Directorate, London, 1998)
17. R. Frank, M.A. Andresen, C. Cheng, P.L. Brantingham, Finding criminal attractors based on offenders' directionality of crimes, in *Proceedings of the 2011 European Intelligence and Security Informatics Conference (EISIC'11)*, 2011, pp. 86–93
18. R. Frank, M.A. Andresen, P.L. Brantingham, Criminal directionality and the structure of urban form. J. Environ. Psychol. **32**(1), 37–42 (2012)
19. R. Frank, M.A. Andresen, P.L. Brantingham, Visualizing the directional bias in property crime incidents for five canadian municipalities. Can. Geogr./Le Géographe canadien **57**(1), 31–42 (2013)
20. R. Frank, B. Kinney, How many ways do offenders travel – evaluating the activity paths of offenders, in *Proceedings of the 2012 European Intelligence and Security Informatics Conference (EISIC'12)*, 2012, pp. 99–106
21. R.G. Golledge, Misconceptions, misinterpretations, and misrepresentations of behavioral approaches in human geography. Environ. Plan. A **13**(11), 1325–1344 (1981)
22. M.C. Gonzalez, C.A. Hidalgo, A. Barabasi, Understanding individual human mobility patterns. Nature **453**(7196), 779–782 (2008)
23. W. Gorr, R. Harries, Introduction to crime forecasting. Int. J. Forecast. **19**(4), 551–555 (2003)
24. G. Grimmett, D. Stirzaker, *Probability and Random Processes* (Oxford University Press, Oxford, 1989)
25. K. Harries, *Mapping Crime Principle and Practice* (U.S. Department of Justice, Office of Justice Programs, National Institute of Justice, Washington, DC, 1999)
26. S.D. Johnson, Repeat burglary victimisation: a tale of two theories. J. Exp. Criminol. **4**(3), 215–240 (2008)
27. S.D. Johnson, A brief history of the analysis of crime concentration. Eur. J. Appl. Math. **21**(4–5), 349–370 (2010)
28. S.D. Johnson, W. Bernasco, K.J. Bowers, H. Elffers, J. Ratcliffe, G. Rengert, M. Townsley, Space–time patterns of risk: A cross national assessment of residential burglary victimization. J. Quant. Criminol. **23**(3), 201–219 (2007)
29. T. Kolokolnikov, M. Ward, J. Wei, The stability of steady-state hot-spot patterns for a reaction-diffusion model of urban crime. Discrete Contin. Dyn. Syst. **34**, 1373–1410 (2014)
30. H. Liu, D.E. Brown. Criminal incident prediction using a point-pattern-based density model. Int. J. Forecast. **19**(4), 603–622 (2003)

31. B. Liu, Y. Fu, Z. Yao, H. Xiong, Learning geographical preferences for point-of-interest recommendation, in *Proceedings of the 19th ACM SIGKDD International Conference on Knowledge Discovery and Data Mining (KDD'13)*, 2013, pp. 1043–1051
32. B.N. Miller, J.A. Konstan, J. Riedl, Pocketlens: toward a personal recommender system. ACM Trans. Inf. Syst. **22**(3), 437–476 (2004)
33. L.A.J. Quetelet, *A Treatise on Man and the Development of His Faculties* (W. and R. Chambers: Edinburgh, 1842)
34. N. Rodriguez, A. Bertozzi, Local existence and uniqueness of solutions to a PDE model for criminal behavior. Math. Models Methods Appl. Sci. **20**(supp01), 1425–1457 (2010)
35. D.K. Rossmo, *Geographic Profiling* (CRC Press, Boca Raton, FL, 2000)
36. D.R. Schaefer, Youth co-offending networks: An investigation of social and spatial effects. Soc. Networks **34**(1), 141–149 (2012)
37. L.W. Sherman, P.R. Gartin, M.E. Buerger, Hot spots of predatory crime: routine activities and the criminology of place. Criminology **27**(1), 27–56 (1989)
38. M.B. Short, A.L. Bertozzi, P.J. Brantingham, Nonlinear patterns in urban crime: hotspots, bifurcations, and suppression. SIAM J. Appl. Dyn. Syst. **9**(2), 462–483 (2010)
39. M.B. Short, P.J. Brantingham, A.L. Bertozzi, G.E. Tita, Dissipation and displacement of hotspots in reaction-diffusion models of crime. Proc. Natl. Acad. Sci. **107**(9), 3961–3965 (2010)
40. M.B. Short, M.R. D'orsogna, V.B. Pasour, G.E. Tita, P.J. Brantingham, A.L. Bertozzi, L.B. Chayes, A statistical model of criminal behavior. Math. Models Methods Appl. Sci. **18**(supp01), 1249–1267 (2008)
41. J. Song, V. Spicer, P.L. Brantingham, R. Frank, Crime ridges: exploring the relationship between crime attractors and offender movement. in *Proceedings of the 2013 European Intelligence And Security Informatics Conference (EISIC'13)* (IEEE, New York, 2013), pp. 75–82
42. M.A. Tayebi, R. Frank, U. Glässer, Understanding the link between social and spatial distance in the crime world, in *Proceedings of the 20nd ACM SIGSPATIAL International Conference on Advances in Geographic Information Systems (ACM SIGSPATIAL GIS'12)*, 2012, pp. 550–553
43. M.A. Tayebi, M. Ester, U. Glässer, P.L. Brantingham, CRIMETRACER: activity space based crime location prediction, in *Proceedings of the 2014 International Conference on Advances in Social Networks Analysis and Mining (ASONAM'14)*, 2014, pp. 472–480
44. M.A. Tayebi, U. Glässer, P.L. Brantingham, Learning where to inspect: location learning for crime prediction, in *Proceedings of the 2015 International Conference on Intelligence and Security Informatics (ISI'15)*, 2015, pp. 25–30
45. M.A. Tayebi, U. Glässer, M. Ester, P.L. Brantingham, Personalized crime location prediction. Eur. J. Appl. Math. **27**, 422–450 (2016)
46. H. Tong, C. Faloutsos, J. Pan, Fast random walk with restart and its applications, in *Proceedings of the 6th International Conference on Data Mining (ICDM'06)*, 2006, pp. 613–622
47. M. Townsley, A. Sidebottom, All offenders are equal but some are more equal than others : variation in journeys to crime between offenders. Criminology **48**(3), 897–917 (2010)
48. United Nations Settlements Programme: The State of the World's Cities 2004/2005: Globalization and Urban Culture (2004), Retrieved from http://mirror.unhabitat.org/pmss/getElectronicVersion.aspx?nr=1163&alt=1
49. H. Wang, M. Terrovitis, N. Mamoulis, Location recommendation in location-based social networks using user check-in data, in *Proceedings of the 21st ACM SIGSPATIAL International Conference on Advances in Geographic Information Systems (ACM SIGSPATIAL GIS'13)*, 2013, pp. 374–383
50. D.L. Weisburd, E.R. Groff, S.-M. Yang, *The Criminology of Place: Street Segments and Our Understanding of the Crime Problem* (Oxford University Press, Oxford, 2012)
51. J.Q. Wilson, G.L. Kelling, Broken windows and police and neighborhood safety. Atlantic **249**, 29–38 (1982)
52. M. Ye, P. Yin, W. Lee, Location recommendation for location-based social networks, in *Proceedings of the 18th ACM SIGSPATIAL International Conference on Advances in Geographic Information Systems (ACM SIGSPATIAL GIS'10)*, 2010, pp. 458–461

Chapter 8
Concluding Remarks

This book extends and integrates multidisciplinary research into a methodological framework for employing social network analysis in predictive policing, an emerging field with high potential to serve as a powerful tool for crime reduction and prevention. Predictive policing enables law enforcement agencies to be smart and effective in deploying their resources. Social network analysis can play a pivotal role in predictive policing by mining patterns of relationship among offenders.

This research covers major problems in predictive policing that can take advantage of social network analysis, and is the first comprehensive work in this domain, to the best of our knowledge. We believe the systematic approach for studying the criminological problems presented in this work opens the door for researchers in criminology and computer science fields to explore important issues pertaining to public safety, and facilitates more informed and deliberate adoption of predictive policing as a complement for existing policing methods.

Our data mining approaches presented throughout this book show that the structure of co-offending networks can provide valuable information for understanding crime patterns and criminal behaviours. The proposed methods extract co-offending patterns embedded in co-offending networks in the node, group and network levels to predict criminal activities. We use the extracted patterns to disrupt co-offending networks, detect organized crime groups, investigate suspects, predict co-offences and predict crime locations in a personalized manner. While crime prediction is one of the most difficult predictive tasks because of complicated patterns behind criminal behaviours, our proposed methods yield high-quality results.

CRIMEWALKER uses partial knowledge of the offenders involved in a crime incident and the structure of a known co-offending network to recommend the top-N potential suspects. CRIMEWALKER extends the existing random walk based models to address link prediction combined with the ability to perform recommendations based on a set of offenders given as input instead of a single offender.

Our supervised learning framework for the co-offence prediction problem covers essential aspects of this problem including strong prediction features extraction

M.A. Tayebi, U. Glässer, *Social Network Analysis in Predictive Policing*,
Lecture Notes in Social Networks, DOI 10.1007/978-3-319-41492-8_8

to increase classification performance and appropriate prediction space definition to overcome the class imbalance problem. By applying different classifiers to the defined prediction spaces our proposed method can correctly predict up to 90 % of all co-offences in the best case scenario.

Despite the importance of organized crime for law enforcement agencies, as far as we know, there is no computational framework for detecting organized crime activities from crime data. Based on Canadian Criminal Code we formalized the concept of organized crime group. Then, we proposed a computational approach using social network analysis prescriptive to adopt a community detection method for detecting organized crime groups. Our proposed approach provides important insights into the ways in which co-offending networks shape and affect criminal behaviour.

CRIMETRACER is a random walk based model which is personalized to predict the crime location of every offender. Our experimental evaluation shows that personalization using co-offending network information contributes in detecting crime locations. Considering the difficulties and importance of modeling offenders' spatial behaviour, and prediction strength of CRIMETRACER which is up to 23 %, we believe that this research is groundbreaking in the spatial crime data mining area.

Multidisciplinary research is challenging, both to the researchers regarding their perspectives and methodologies and to policymakers and practitioners regarding the way the 'problem' is defined. In this research we have collaborated closely with criminologists throughout all phases from problem definition to solution design to result analysis. For our experimental evaluation we use real-world crime data, and major parts of this research are carried out in close collaboration with experts in law enforcement agencies; for instance, the work presented in the section on organized crime group detection is the result of a project defined and conducted by Public Safety Canada, Organized Crime Division and RCMP "E" Division.

Predictive policing as a multi-step process has important operational challenges. Three main steps of this process, question formulation, data preparation, and data mining, have been studied in depth in this research. The ultimate steps, police acting on and efficacy evaluation, are beyond the scope of the research presented here. Predictive policing that can make a difference needs an iterative process where law enforcement and policymakers act on analytics derived through crime data mining at the strategic, the tactical, and the operational level.

We believe that the ideas presented here can inspire new research directions in social network analysis and data mining with useful applications for predictive policing, criminal investigations, and criminal intelligence in the endeavor to combat crime. However, many open problems still remain in the realm of predictive policing for taking full advantage of social network analysis. Here, we briefly outline several potential future research directions in the field of predictive policing.

- **Improving the proposed methods.** There are a number of possibilities to enhance the performance of the proposed methods. For instance, in CRIMETRACER enhancing movement directionality element can contribute to the method performance. A suggestion is to personalize hotsposts for every offender instead of using all hotspots. In the proposed supervised co-offence prediction framework counterintuitively time-based features work weaker than the original version of the corresponding feature. Parameter regularization of these features for improving their prediction strength is an extension of this work. For detecting organized crime groups and co-offending network key players we do not consider geographic factors. An interesting research direction to improve the proposed methods is taking geographic factors into account.

- **Mining multiple related criminal networks.** All proposed models in this book take one co-offending network extracted from police-arrest data as input. As discussed before, because of the nature of crime and criminal behaviour crime datasets are incomplete. Therefore, integrating different resources to have a more complete picture of offenders' activities is essential. For instance, in detecting organized crime groups intelligence service data can be used as complement of police-arrest data. An interesting direction to extend this book is using multiple criminal networks for predictive policing tasks.
- **Learning from collective human behaviour.** Mobile phone data provide rich information on population movement in urban area which makes it possible the study of collective human behaviour. In [2], the authors use mobile phone data to predict crime hotspots. Enriching the personalized crime location prediction and other predictive policing methods using data sources such as mobile phone data is a possibility for future work.
- **Organized crime group evolution.** For tracking organized crime group evolution we apply a matching function on detected groups over a number of time steps. The other direction is using evolutionary graph clustering [5] for studying evolution of organized crime groups.
- **Predicting crime location of multiple co-offenders.** CRIMETRACER can be extended to predict the location of crimes committed by multiple co-offenders. Such a method should use the activity space of individual offenders generated by CRIMETRACER to learn the locations where a group of co-offenders commits crimes. This research will address an important criminological question about how activity space of an offender can be influenced by his co-offenders' activity space.
- **Detecting criminal groups' activity space.** Existing works for criminal group's activity space detection [1, 3] focus only on situations with two groups, and only find the boundaries between groups' territories. These models neither differentiate between crime types, nor consider if a crime is committed by an offender or a group of offenders. Another future research direction is mining co-location patterns for detecting criminal groups' activity space.
- **Minimizing the epidemic spread of undesirable behaviours.** The influence of social interactions in forming people behaviours is now widely studied and recognized in many areas. This is true in the crime world, where the lack of formal education for criminal skills plays an important role in the formation of criminals' behaviours and criminal networks. For performing criminal acts, obviously learning illegal behaviours must depend on informal networks and

peer-to peer contacts, because there is no formal type of learning to become a criminal. According to [9], criminal behaviour is the result of learning an "excess of definitions favorable to law violation." Later different studies discuss how criminal activities may benefit from social interactions by sharing proper know-how about crime business [4], by recruiting young criminals [8] or by transferring skills [6]. A possible research problem is minimizing the spread of undesirable criminal behaviours in a co-offending network assuming the independent cascade model [7] as a mathematical model of behaviour diffusion.

References

1. A.B.T. Barbaro, L. Chayes, M.R. D'Orsogna, Territorial development based on graffiti: a statistical mechanics approach. Phys. A. Stat. Mech. Appl. **392**(1), 252–270 (2013)
2. A. Bogomolov, B. Lepri, J. Staiano, N. Oliver, F. Pianesi, A. Pentland, Once upon a crime: towards crime prediction from demographics and mobile data, In *Proceedings of the 16th International Conference on Multimodal Interaction (ICMI'14)* (2014), pp. 427–434
3. P.J. Brantingham, G. Tita, M.B. Short, S. Reid, The ecology of gang territorial boundaries. Criminology **50**(3), 851–885 (2012)
4. A. Calvó-Armengol, Y. Zenou, Social networks and crime decisions: the role of social structure in facilitating delinquent behavior. Int. Econ. Rev. **45**(3), 939–958 (2004)
5. D. Chakrabarti, R. Kumar, A. Tomkins, Evolutionary clustering, in *Proceedings of the 12th ACM SIGKDD International Conference on Knowledge Discovery and Data Mining (KDD'06)* (2006), pp. 554–560
6. E.L. Glaeser, B. Sacerdote, J.A. Scheinkman, Crime and social interactions. Technical Report, National Bureau of Economic Research, 1995
7. D. Kempe, J. Kleinberg, É. Tardos, Maximizing the spread of influence through a social network, in *Proceedings of the 9th ACM SIGKDD International Conference on Knowledge Discovery and Data Mining (KDD'03)* (2003), pp. 591–600
8. A.J. Reiss Jr., Co-offending and criminal careers. Crime Justice **10**, 117–170 (1988)
9. E.H. Sutherland, *Principles of Criminology* (J. B. Lippincott & Co., Chicago, 1947)

Index

A
active offender group, 45
activity node, 81
activity path, 81
activity space, 81, 100, 102
anchor location, 105
Apriori, 71
association rule mining, 71
association rules, 71
attributed hypergraph, 16
AUC, 90
average distance, 22

B
BC crime dataset, 18
betweenness centrality, 29
BFS algorithm, 22
broken window, 8

C
centrality analysis, 10
centrality measure, 28
chi-square, 69
class imbalance, 85
clique, 41
clique percolation method, 39
closeness centrality, 29
clustering coefficient, 23
co-authorship network, 70
co-offence prediction, 13, 83
co-offending network disruption, 12
co-offending network model, 17
coldspot, 99
collaborative filtering, 105
common activity space, 81
community detection, 10, 24, 42
community evolution tracking, 43
community policing, 8
computational criminology, 2
confidence, 71
connected component, 24
connecting path, 22
crime, 1
CRIMETRACER, 106
crime analysis, 1, 7
crime attractor, 103
crime data, 15
crime data model, 16
crime generator, 103
crime location prediction, 13, 104
crime occurrence space, 102
crime pattern theory, 81, 102
CrimeWalker, 65
criminal cooperation opportunity, 84
cross validation, 92

D
data science, 1
DBLP dataset, 70
degree centrality, 28
degree distribution, 20
differential association, 84
Dijkstra's algorithm, 22
directionality, 103
disorder policing, 8
drug crime, 19

M.A. Tayebi, U. Glässer, *Social Network Analysis in Predictive Policing*,
Lecture Notes in Social Networks, DOI 10.1007/978-3-319-41492-8

E
edge betweenness, 24
effective diameter, 23
eigenvector centrality, 29
ensemble method, 92
environmental criminology, 102
Erdos-Renyi model, 20
evolution model, 48
evolutionary clustering, 43
experience-related, 84

F
FP-growth, 71

G
generative models, 10
geo-social feature, 87
geographic feature, 87
geographic profiling, 81
geographically-related, 84
Girvan-Newman algorithm, 24
group activity, 45
group criminality, 45

H
homophily, 68, 89
hotspot, 99
hotspots influence, 110
hotspots policing, 8

I
information diffusion, 10
intermediate anchor locations, 109

J
Jaccard index, 33

K
key player, 28

L
large components, 24
link prediction, 10

M
main anchor locations, 109
matching function, 48
maximum likelihood, 20
medium components, 24
modularity, 43
moral crime, 19
movement directionality, 110

N
network diameter, 23
network evolution, 25
NP complete, 43

O
o-offending network, 10
offender group, 44
offender trend, 105
organized crime, 40
organized crime group detection, 12, 45

P
PageRank, 30
PIRS, 18, 58
Poisson distribution, 20
police administration, 8
power law, 20
precision, 114
prediction feature, 87
predictive policing, 7
problem-oriented policing, 8
property crime, 19
public safety, 2
Public Safety Canada, 57

R
random walk, 65, 106, 107, 115
RCMP, 18
recall, 114
repeat offender, 114
road feature, 105
road feature weight, 110
road network, 105
road segment, 105
ROC, 90
RWR, 65, 115

S
scale free, 20
selection, 68
serious crime, 19
serious offender group, 45
similarity feature, 89

small sized components, 24
SNA, 9
social feature, 87
social influence, 68
social network, 9
social network analysis, 9
socially-related, 84
standard model of policing, 8
strategic crime analysis, 1
strength, 18, 20
supervised random walks, 110
support, 71
suspect investigation, 13, 64

T
tactical crime analysis, 1
temporal smoothness, 43
top-N recommendation, 66, 114
transaction, 71
transience, 33
transitivity, 23

U
utility, 114

V
visual analysis, 10

W
Watts and Strogatz model, 20
Weka, 90

Printed in the United States
By Bookmasters

Printed in the United States
By Bookmasters